PROPERTY OF:
DAVID O. McKAY LIBRARY
IDAHO
C405

D0214605

FEB 13 2009

WITHDRAWN

OCT 1 4 2022

DAVID O. McKAY LIBRARY
BYU-IDAHO

# Earth Science and Human History 101

# EARTH SCIENCE AND HUMAN HISTORY 101

JOHN J.W. ROGERS AND TRILEIGH TUCKER

SCIENCE 101

**GREENWOOD PRESS**
Westport, Connecticut • London

**Library of Congress Cataloging-in-Publication Data**

Rogers, John J. W. (John James William), 1930–
   Earth science and human history 101 / John J.W. Rogers and Trileigh Tucker.
     p.   cm. — (Science 101, ISSN 1931–3950)
   Includes bibliographical references and index.
   ISBN 978–0–313–35558–5 (alk. paper)
   1.  Historical geology.   2.  Earth sciences—History.   3.  Nature and civilization.
I.  Tucker, Trileigh, 1955–   II.  Title.   III.  Title: Earth science and human history
one hundred and one.
   QE28.3.R63   2008
   304.2—dc22       2008016571

British Library Cataloguing in Publication Data is available.

Copyright © 2008 by John J.W. Rogers and Trileigh Tucker

All rights reserved. No portion of this book may be
reproduced, by any process or technique, without the
express written consent of the publisher.

Library of Congress Catalog Card Number: 2008016571
ISBN-13: 978–0–313–35558–5
ISSN: 1931–3950

First published in 2008

Greenwood Press, 88 Post Road West, Westport, CT 06881
An imprint of Greenwood Publishing Group, Inc.
www.greenwood.com

Printed in the United States of America

The paper used in this book complies with the
Permanent Paper Standard issued by the National
Information Standards Organization (Z39.48–1984).

10 9 8 7 6 5 4 3 2 1

# CONTENTS

# SERIES FOREWORD

What should you know about science? Because science is so central to life in the 21st century, science educators believe that it is essential that *everyone* understand the basic foundations of the most vital and far-reaching scientific disciplines. *Earth Science and Human History 101* helps you reach that goal—this series provides readers of all abilities with an accessible summary of the ideas, people, and impacts of major fields of scientific research. The volumes in the series provide readers—whether students new to the science or just interested members of the lay public—with the essentials of a science using a minimum of jargon and mathematics. In each volume, more complicated ideas build upon simpler ones, and concepts are discussed in short, concise segments that make them more easily understood. In addition, each volume provides an easy-to-use glossary and an annotated bibliography of the most useful and accessible print and electronic resources that are currently available.

# PREFACE

Not only do we live on the earth, but we must also live with it; we can live better with the earth if we understand it, know something about the processes that formed the earth and are still changing it now, and understand how these processes affected human history and may change that history in the future.

This book explores the relationship between the earth and people. We discuss how people have always lived by taking resources from the earth and returning their waste products to it. In addition to the air that we breathe and the fresh water that we drink, people require rock and ore to make tools and buildings, food to eat, and a source of energy.

The types of materials that people require have changed remarkably throughout the history of the human race. We can follow those developments by a combination of archaeological information and, more recently, by written records. Very early people hunted and foraged for food, used only rocks for tools, and made fires with wood. We trace the gradual progression from this lifestyle to the complexities and advantages of modern life.

People now have an enormous variety and abundance of foods to eat, energy sources that range from fossil fuels to nuclear and renewable energy, and materials that range from rock to special steel alloys. We discuss how people obtain food, energy, and materials today. This discussion leads to questions about future availability and whether we can continue to live as well as we do today.

The earth is constantly changing and the changes affect people in many ways in addition to providing the necessities of life. Some of the changes are slow, but some are rapid and violent. Slow shifts in river courses can affect national boundaries and trade routes, and floods can overwhelm people. Normal wind patterns distribute moisture around the earth and have always affected transportation, but occasional hurricanes can be disasters. Slow movement along faults contributes to the movement of tectonic plates, but rapid fracturing causes earthquakes, some of which generate tsunamis. Slow volcanic eruptions have covered much of the earth's surface with lava and ash, but violent eruptions can bury cities and put so much ash into the atmosphere that worldwide cooling occurs.

We pay particular attention to present controversies about the earth and its effects on people. What are the causes and consequences of global warming, and what, if anything, can we do about them? Are rich societies prospering by putting extra hardships on poor people? Are biblical accounts of creation and the Noachian flood and the prophecy of the battle of Armageddon, accurate? Conversely, can we explain the origin of animals and plants by evolution, the flood as a natural occurrence in the Black Sea, and Armageddon as an earthquake?

American historian Will Durant is reported, perhaps erroneously, to have said: "Civilization exists by geological consent, subject to change without notice." We agree.

# INTRODUCTION

Any effort to describe the relationship between the earth and people's lives requires arbitrary decisions. We divide the discussion into four major sections, each of which cross-references other sections where necessary. The chapters are:

1. Atmosphere, Oceans, and Rivers
2. Tectonics
3. Evolution, Creationism, and the Long History of the Earth
4. Resources and the Environment

Chapter 1 covers processes that occur on and above the earth's surface. The first part shows how the composition of the atmosphere controls the earth's climate. High-energy radiation from the sun passes through the atmosphere to the earth's surface and then is radiated back from the surface as low-energy infrared waves. Because greenhouse gases in the atmosphere absorb this infrared radiation, the earth's surface is at a temperature that permits animals and plants to live, but the present increase in the concentration of these gases is contributing to global warming.

Variation in the intensity of solar radiation occurs over several time scales. Cycles that lasted tens of thousands of years caused repeated glaciations and interglacial intervals in the last 400,000 years. The last glacial period ended rapidly about 15,000 years ago and the earth may be about to enter a new period of glaciation.

Irradiance variations that lasted a few hundred years have strongly affected human history in the past 2,000 years. The Roman Empire expanded during a period of high temperatures and was destroyed during a following period of cold years. The population of America expanded as people emigrated from Europe during the coldest part of the Little Ice Age, from the 17th to the 19th centuries.

Present global warming is caused partly by increase in solar irradiance, but increase in greenhouse gases by human activities is probably a more important reason for warming. We discuss the consequences of warming, the possibility of reducing it by changes in our use of energy, and the speed with which climate changes occur.

Enormous volcanic eruptions can put so much ash in the atmosphere that they cause global cooling for many years. The two eruptions that had the greatest effect on human history were probably both in Indonesia. Toba blew up about 75,000 years ago with such force that the atmospheric cooling it caused might have killed so many people that small bands of survivors became inbred and differentiated into the diversity of modern human races. The eruption of Krakatoa in 535 AD may have unleashed the first episode of plague (Black Death) on the world.

Both winds and ocean currents (gyres) are caused by rotation of the earth. Trade winds (easterlies) blow toward the west in the tropics, and westerlies blow toward the east in temperate regions. These wind patterns distribute moisture around the earth and affect the flight times of modern airplanes. The basic (Coriolis) effect that creates these wind patterns also causes rotation of hurricanes, and we discuss Hurricane Katrina in 2005 as an example.

Gyres distribute heat and nutrients in the oceans and affected travel times of old sailing ships. Ocean temperatures can also be affected by other processes. We discuss the phenomenon known as El Nino and how it may have destroyed the Moche civilization of Peru.

Rivers and their drainage basins change continually. Some of the changes are seasonal floods, but many modifications are longer lasting as channels shift and basins become larger or smaller. Whether short-term or long-term, all changes in rivers affect the lives of people who live near them. The annual flooding of the Nile River controlled the lives of Egyptians. The Mississippi River would divert its flow to the Atchafalaya River and abandon New Orleans if the U.S. Corps of Engineers didn't prevent the river diversion. We also discuss how the biblical (Noachian) flood may have occurred in the Black Sea instead of along rivers in the Middle East.

The final discussion in Chapter 1 is about the importance of canals for human history. It concentrates on the Suez and Panama Canals.

Chapter 2 considers the nature of the solid earth and the processes that occur within it. We discuss the history of the development of our understanding of the earth in order to give readers an insight into the problems that geologists face.

The three major zones of the earth—core, mantle, and crust—were discovered largely by seismic methods. Combining knowledge of the mantle and crust with the principle of isostasy that had been discovered earlier led to the understanding that mountains are high because they have deep roots of light rock that extend down into the mantle. This combined knowledge also explained that continents float higher than ocean basins because continental crust is lighter and thicker than oceanic crust.

One of the problems faced by early geologists was the amount of heat in the earth. Before the discovery of radioactivity, geologists assumed the earth was cooling down and formed mountains by shrinking of the crust. A source of heat by radioactive decay, however, led geologists to look for other ways to explain the earth's tectonic processes.

Geologists discovered these processes throughout the 20th century. They started with the different types of faults that had been established in the 19th century, added the concept of continental rifting, the likelihood of continental drift, and subduction along some continental margins and all island arcs. When sea floor spreading was discovered in the early 1960s, geologists had all the information they needed to develop plate tectonics.

Plate tectonics is the concept that the earth's surface is divided into stable plates separated by different types of active plate margins. The plate margins include: subduction zones, where compression occurs; mid-ocean ridges, where spreading occurs, and transform margins, where plates move laterally past each other.

Hot spots are mysterious phenomena that bear little or no relationship to plate tectonics. They are places where volcanism continues for tens of millions of years as plates move over them and carry their older products away from the active spot. They may be caused by plumes that rise from the mantle, and we discuss Hawaii and Midway Island as a type example.

The second part of Chapter 2 describes the relationships between tectonics and people's lives. Earthquakes are one consequence of plate movements. The 1906 earthquake in San Francisco taught geologists and engineers better ways to design buildings and infrastructure. It is also possible that the biblical battle of Armageddon was prophesied on the basis of known earthquakes at the Israeli site of Har Megiddo. The earthquake near Sumatra in 2004 generated a tsunami that killed tens of thousands of people.

Volcanoes are another consequence of plate movements. We discuss eruptions that altered climate in Chapter 1, and here we describe eruptions that had mostly local effects. One of the most significant was the eruption of Santorini in the Mediterranean in 1640 BC. That eruption may have ended Minoan dominance in the area and led to its replacement by Mycenaeans.

Rift valleys have always been favorable places to live. The human race may have evolved in the East African rift system, and modern Icelanders enjoy the hot water provided by the mid-ocean ridges that cross Iceland.

Small islands pose numerous problems for people who live on them. We concentrate on Hawaii and Midway Island, which were important landing points in trans-Pacific aviation before development of modern planes that could fly across the Pacific Ocean without stopping. Those landing points were major targets of the Japanese when they attacked Pearl Harbor and, later, during the battle of Midway in World War II.

Chapter 3 describes controversies about evolution, creationism, and the long history of the earth. Creationism is the biblical idea that God specially created all animals and plants, and many creationists believe that this event occurred when the earth was created in 4004 BC.

Creationism has always been somewhat controversial. The disputes intensified, however, when geologists developed the concept of uniformitarianism and biologists proposed evolution. The disagreements were brought sharply into focus with the Scopes trial in 1925 and continue today in American school boards and the court system.

Charles Darwin and his supporters originally proposed evolution on the basis of observations of modern organisms. Paleontologists are now able to support that concept by showing how life developed during the history of the earth. We concentrate on the first appearance of various types of organisms, including cyanobacteria, multicellular algae, animals, vertebrates, land plants, and mammals.

The paleontologic record also contains at least three (possibly four) extinction events. They are times when old forms of organisms went extinct and were replaced by new forms. The most famous extinction event occurred when a large asteroid struck the earth about 65 million years ago and finished off the dinosaurs. Even larger extinctions probably occurred 250 million years ago and 540 million years ago. We also discuss the possibility that large animals (megafauna) were killed by overhunting in the past few tens of thousands of years as people moved into formerly unoccupied lands.

Geologists began in the 19th century to establish a relative sequence of events in the history of the earth. This time scale made it possible for paleontologists to work out the sequence of fossils that supported the concept of evolution, but it did not provide any absolute ages of events in the history of the earth. Those ages could not be assigned until radiometric dating methods were developed in the 20th century.

Three of the most important radiometric methods are: the U-Pb system, which is used to date very old rocks; the K-Ar system, which is particularly important to archaeologists who study the last few million years; and radiocarbon (carbon-14) which provides dates for about the past 20,000 years. Somewhat more elaborate methods were needed to establish the formation of the earth at about 4.5 billion years ago.

The long history of the earth is preserved in continents because the oldest rock in ocean basins is no more than about 200 million years old. Continental rock began to develop shortly after the earth was formed, but the oldest true continents are only about 3 billion years old. We discuss the formation of continents and also the history of supercontinents.

Supercontinents were assemblages of all of the world's continental crust into one landmass. The most famous was Pangea, which existed about 250 million years ago. It was preceded by Gondwana, with an age of about 500 million years, Rodinia with an age of 1 billion years, and Columbia at an age between 1.6 and 1.8 billion years. This sequence requires some process in the mantle that causes alternate cycles of accretion and dispersal.

Chapter 4 shows how people obtain the requirements for life from the earth. These requirements include: sources of energy; rocks, and other raw materials; food; water; and a resource that has become extremely important in the last few years—wireless communication and the Internet.

The first part of the discussion of energy is a comparison of the consumption and production of different types of energy in the United States and the rest of the world. This information makes it clear that the United States and other rich countries have much higher per capita consumption than poor countries.

The most important modern energy source is fossil fuel, including oil, coal, and natural gas. We describe the formation of these types of fuels and then discuss the history of their use.

Early humans used wood fires as their only way of producing heat and light and water wheels as a minor source of energy. Sources of energy became slightly more complicated when animal and vegetable oils were used for lamps and tallow was used for candles. The industrial revolution that began in the 18th century was powered mostly by coal. New types of machinery, such as steam engines and electric motors, were soon invented to make industry even more productive. Oil was first produced in the 19th century and has now become the principal source of energy in the United States and the rest of the world.

The United States was self-sufficient in energy until about 1970, but then it began to import oil. Other sources of energy became important as both the volume of imports and the price of oil increased. These sources include nuclear energy, which many people regard as hazardous, and renewable sources such as solar power, hydropower, wind power, and biofuels.

Progressive development of the raw materials that people use took place more rapidly than the development of fuels. Early people used only rock, but copper began to be smelted by about 4000 BC. The Copper Age soon gave way to the Bronze Age and then the Iron Age. People now use a variety of metals, such as aluminum and lead, and they produce numerous types of steel.

The use of nonmetals also changed through time. The ones most abundantly used now are cement, concrete (cement plus gravel), and fertilizer. Precious metals and gems have always been important, and we discuss gold and diamonds.

Sources of food have grown more widespread as history progressed. Early humans lived by hunting animals and gathering food from the forest. People continued to live on food produced locally even when agriculture developed in the general time range of 10,000–5000 BC. Wider distribution of different types of food occurred when Europeans began arriving in the Americas in the 16th and 17th centuries. The agricultural exchange brought maize (corn), beans, and potatoes from the Americas to the rest of the world and brought domestic animals and wheat to the Americas. Now different kinds of food are traded internationally, and local cuisines are less important than they used to be.

Inequalities in the consumption of food and water are similar to those in the consumption of energy. Industrial societies use enormous amounts of water, and rich people in all societies consume larger quantities of food than poor people. Diets of rich people and societies, however, are not always healthy, and we demonstrate the problems by a discussion of nutritional requirements.

Communication is now one of the world's most important resources. It has progressed from the tedious, 2-year, trip along the Silk Road between China and Europe to the instantaneous Internet. Intermediate stages include faster travel on land and sea, telegraphy, telephones, and radio. We conclude by showing the importance of communication to people's freedom.

# 1

# ATMOSPHERE, OCEANS, AND RIVERS

Everybody has to breathe air and drink fresh water. Beyond those requirements, our lives are affected in broader ways by the atmosphere, oceans, and rivers. We explore those effects in this section, recognizing how people responded to them in the past and how they may change our lives in the future.

We start with a discussion of the composition of the atmosphere. In addition to the oxygen that animals need and the nitrogen that plants need, the atmosphere contains much smaller amounts of "greenhouse gases" such as water vapor, carbon dioxide, and methane. It also contains tiny concentrations of ozone, which have an important effect on ultraviolet radiation.

Understanding the composition of the atmosphere leads to a discussion of the processes that control the earth's climate. They include not only the composition of the atmosphere but also changes in the amount of solar radiation that reaches the earth. These changes have been large enough in the past four hundred thousand years to cause repeated cycles of glaciation and deglaciation. We describe how the most recent cycle affected the arrival of the first people in the Americas and may have been responsible for stories of the biblical (Noachian) flood.

Small changes in climate in the last 2,000 years caused people to move around the earth to find better places to live. The Roman Empire expanded when the climate was warm and contracted when temperatures fell. Agriculture flourished in Europe during the warm temperatures of the Middle Ages and contracted when the Little Ice Age began in the 14th century. The skyrocketing price of food in Europe sent many people to America and helped establish the country as a world power.

Increase in concentrations of greenhouse gases in the atmosphere, and possibly increase in solar radiation, are now causing rapid global warming. It is already affecting life in many parts of the world, including Alaska, Greenland, and northern Africa. We discuss the consequences of abrupt changes in the future, including drastic cooling of North America and Europe.

Few people understand the dramatic effects that volcanic eruptions have on the earth's climate because of cooling caused by ash in the atmosphere. Most of them have affected people for only a few years, but at least two may have changed the history of the human race. Three eruptions in Indonesia are particularly important: Toba in 73,000 BC, Krakatoa in 535 AD, and Tambora in 1815. In addition, we discuss the effects of an eruption in Iceland in 1783 and how eruptions in Peru and the Solomon Islands may have affected English colonies at Jamestown and the Lost Colony.

We must also consider the consequences of major volcanic eruptions in the future. Long-term predictions are impossible, but potential dangers can be recognized. One of those dangers is another eruption of Yellowstone, which could have climatic effects as well as blanketing much of the western United States under ash.

Interaction of the rotation of the earth with the atmosphere and oceans creates wind and ocean currents (gyres). Sailing ships plotted routes to take advantage of both wind and currents, and prevailing winds affect the flight times of modern airplanes. We discuss how wind and currents affected the route of the Manila Galleon when the Spanish controlled the Pacific Ocean and also how the phenomenon known as El Nino contributed to the destruction of the Moche Empire of Peru.

Some of the movements of air and water are violent. About 10 hurricanes reach the United States each year, and we discuss Hurricane Katrina in 2005 as an example. We defer the discussion of the Indian Ocean tsunami of 2004 to Chapter 2, however, because it was caused by an earthquake.

River systems change continually over time. Rivers cut valleys and move from side to side as they erode at some places and deposit sediment at others. These changes have important consequences for people, and we discuss five. The Lewis and Clark expedition explored the Missouri River and found that there is no easy route to the Pacific Ocean. If the Atchafalaya River of Louisiana erodes farther northward, it could drain water out of the Mississippi River and close the port of New Orleans. Changes in the course of the Rio Grande River made it necessary for the United States and Mexico to adjust the international border between El Paso and Juarez. The annual flooding of the Nile River has affected civilizations along its valley throughout history. Settlers moving by wagons to California had to cross the Great Basin, where rivers disappear into the ground and leave a waterless stretch where many animals died.

Canals have been important contributors to history for thousands of years. We concentrate on the Suez and Panama canals.

## ATMOSPHERE COMPOSITION

The earth's atmosphere consists almost entirely of nitrogen (78%) and oxygen (21%). The nitrogen is mostly inert, but oxygen sustains animal life and enters into numerous chemical reactions. One of the most important reactions of oxygen is to combine with iron to form reddish iron oxides ("rust") that give much of the earth's land surface a red color. The most important minor gases vary in abundance, with carbon dioxide averaging about 0.04 percent and water vapor in the range of 0–4 percent.

Without carbon dioxide and water in the atmosphere the temperature of the earth's surface would be similar to that of the moon, nearly the coldness of space. These gases keep the earth's surface warm because of a process known as the "greenhouse effect." Radiation coming from the sun at high energy (high-frequency visible and ultraviolet light) penetrates the atmosphere and is absorbed by the earth's surface. Then the surface radiates the energy back into the atmosphere at lower energy (infrared) waves that are absorbed by the carbon dioxide and water vapor.

This process of warming the atmosphere is called the greenhouse effect because greenhouses operate in an identical way. Their glass passes sunlight through to the interior and then absorbs radiation from the interior as it passes back through the glass toward the outside.

Two gases occur in the atmosphere at very low concentrations but have important effects. One is methane ($CH_4$), which is a greenhouse gas whose concentration has approximately tripled in the last two centuries (see Causes of Global Warming). The second gas is ozone ($O_3$), which has a concentration in the stratosphere of about 10 parts per million (0.001%). We discuss it next.

### Ozone Depletion

The "ozone hole" was first discovered in 1984 by British research workers at the Antarctic research station of Halley Bay, on an ice shelf on the eastern edge of the Weddell Sea. They found that the concentration of ozone in the upper atmosphere (stratosphere) fell to very low levels during each Antarctic spring (September and October). When this observation was confirmed at other research stations and by satellite measurements, people realized that the Antarctic and some areas farther north were covered by an enormous ozone hole as large as North America.

Ozone ($O_3$) develops in the stratosphere when ultraviolet (UV) rays from the sun break normal oxygen ($O_2$) into individual oxygen atoms (O), and the O and $O_2$ combine to form $O_3$. The concentration of $O_3$ normally remains constant because it decomposes back to $O_2$ and O at about the same rate at which it forms. The decomposition can be speeded up, however, when it is "catalyzed" on the surface of solid particles or by other gases. When that happens, the rate of decomposition exceeds the rate of production, and the ozone concentration falls. The catalyst that destroys ozone in the Antarctic is ice crystals that form in the stratosphere when sunlight returns to the Antarctic in the spring.

Ice crystals clearly weren't responsible for ozone reduction that scientists soon found to be occurring at a rate of about 2 percent per decade over the entire earth. The explanation for that reduction was provided by Sherwood Rowland, of the University of California, Irvine, Mario Molina of MIT, and Paul Crutzen of the Max Planck Institute for Chemistry. They discovered that chlorine and bromine ions in the stratosphere catalyzed the breakdown of ozone and were so efficient that one Cl ion could help decompose 100,000 ozone molecules. They shared the Nobel Prize in Chemistry in 1995.

Reduction in the concentration of stratospheric ozone is serious because ozone absorbs the shortest wavelength of ultraviolet radiation (UVB and UVC) and reduces the amount that reaches the earth's surface from the sun. This reduction is essential because UVB/UVC is dangerous, causing a variety of problems ranging from mild skin cancer to more dangerous cancers, cataracts, and other damage to eyes, and even genetic defects.

Investigators soon learned that a principal source of stratospheric chlorine was a group of industrial chemicals known as chlorofluorocarbons (CFCs). These compounds of chlorine, fluorine, and carbon had numerous industrial uses, including cleaning agents, aerosols, and coolants in refrigerators and air conditioners (freon). They also included bromine-containing gases that were used in portable fire extinguishers. Unfortunately, when CFCs reached the stratosphere they were broken down by sunlight, releasing chlorine and, thereby, destroying ozone.

By 1987 most industrial countries signed the Montreal Protocol, which specified that the use of CFCs must be discontinued as soon as possible.

Within a few years after 1987 virtually all CFCs had been replaced by substitutes, including hydrogen-bearing CFCs (HCFCs), which break down so quickly in the lower atmosphere that they do not release chlorine into the stratosphere.

The Montreal Protocol is now regarded as an outstanding example of international cooperation on environmental problems. Early in the 21st century, measurements showed that ozone concentrations in the atmosphere in temperate latitudes had risen back to levels that are regarded as normal for an atmosphere not affected by human activity.

## CHANGES IN SOLAR RADIATION

The amount of sunlight that reaches the earth undergoes cyclic changes. They include cycles that last tens of thousands of years and also variations over a period of a few hundred years. We discuss both types of variation and their consequences for human history

### Long-Term (Tens of Thousands of Years) Cycles

Initial information on long-term cycles came from work at a Russian research station in the Antarctic. (Note how important Antarctic research has been for studies of the atmosphere.)

When America built a base at the South Pole in 1950, it was named the Amundsen-Scott station for the leaders of the two parties that first reached the Pole. When the Soviet Union countered by establishing a research station on the icecap, they named it Vostok, which means "East" in Russian. Vostok is 2,000 feet higher and even colder than the Pole (on July 12, 1983, the temperature at Vostok was a world record –88°C).

The purpose of the research station at Vostok was to drill a core into the underlying ice, and ultimately the small group of people who lived there drilled more than 3 kilometers down into the Antarctic icecap. The drilling stopped when remote sensing showed that the drill was only a short distance above a large body of water, either a lake at the base of the icecap or perhaps an arm of the ocean. The drillers didn't want to take a chance of contaminating this body of water and decided to go no farther.

The cores obtained from Vostok and other places have been invaluable to scientists worldwide. A major reason is that ice traps small bubbles of the atmosphere as it forms. The gas in these bubbles can then be extracted in a laboratory and analyzed to yield the composition of the atmosphere at the time the ice formed. This information throughout the core shows the history of changes in the atmosphere because it is also possible to discover how old the ice is at any depth in the core.

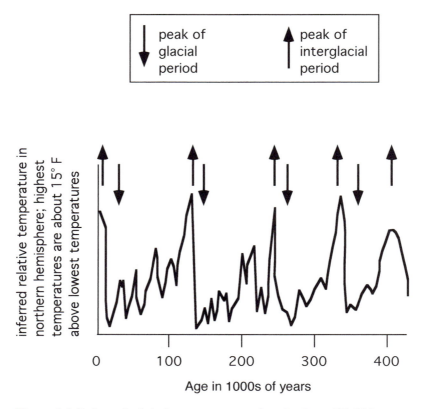

**Figure 1.1** Inferred global temperatures for the last 420,000 years from Vostok and other ice cores.

The simplest way to determine age of the ice is to count annual layers down from the ice surface. Each summer a tiny amount of ice evaporates from the ice surface, and this layer stays in the ice below next winter's accumulation. This count of layers can be confirmed by measuring short-lived radioactive isotopes in the ice. By plotting the age of ice in the core against the composition of trapped gas, we obtain graphs of the type shown in Figure 1.1. It extends back to 420,000 years ago because that was the bottom of the hole at Vostok when drilling stopped.

The measurement of carbon dioxide trapped in the ice provides information about changes in atmospheric temperature. Measurements at several research stations show that carbon dioxide concentrations are high when temperatures are high in the summer because animals are more active and exhale more carbon dioxide into the atmosphere. This relationship appears to be true when averaged over entire years in ice layers, and we use Figure 1.1 as a record of changes in global atmospheric temperatures for the past 420,000 years.

The most prominent variation shown in the graph is an approximate cycle of 100,000 years of slow cooling (glacial periods) and 10,000–20,000 years of rapid warming (interglacial periods). Today we seem to be near the end of an interglacial period that follows the end of the last glaciation 15,000 years ago.

A periodicity of 100,000 years is close to one of the periodic changes in the earth's orbit around the sun. These periodicities are referred to as "Croll–Milankovitch cycles" because they were first recognized by James Croll, a Scottish scientist, and later refined by Milutin Milankovitch, a Serbian scientist. They found a 100,000-year cycle for the "eccentricity" of the earth's orbit, caused by slight changes between a circular and more elliptical shape of the orbit.

Croll and Milankovitch also found two other cycles. With a periodicity of about 20,000 years, the orientation of the earth's rotational axis moves in a small circle, referred to as "precession." This movement shows up well on Figure 1.1 as five smaller peaks in the last 100,000 years. A third orbital cycle known as "obliquity" refers to a "rocking' of the earth's axis back and forth through an angle of about 3°. It has a periodicity of 41,000 years and is hard to find on Figure 1.1. Obliquity variations, however, can be inferred by a mathematical technique known as Fourier analysis.

Most geologists now agree that orbital variations are the "primary" reason for the alternation of glacial and interglacial cycles. They result from the different amounts of sunlight that the earth receives when its axis is in different positions and its path around the sun changes gradually. These variations, however, cannot explain why the earth cools down slowly and warms up rapidly.

**The Last Ice Age and the Consequences of Its End**

The last Ice Age ended about 15,000 years ago. The melting ice uncovered the land and exposed much of the topography of North America and northern Europe. Meltwater filled the oceans, isolating the Americas from the rest of the world and bringing seawater into the Black Sea. We illustrate these consequences with three discussions: topographic effects of deglaciation; peopling of the Americas; and the possibility that the biblical (Noachian) flood occurred in the Black Sea.

*Topographic Effects of Deglaciation*

In the early 19th century, a few geologists proposed that Europe was just emerging from a period of widespread glaciation. This idea was firmly established in 1840 when Louis Agassiz (who was actually an expert on fish) published a definitive work titled "Etudes sur les Glaciers." After this publication, geologists in North America found similar features that could have been developed only in areas formerly covered by ice.

Erosion by former glaciers is shown by valleys where glaciers retreated to the heads of valleys that appeared to have been completely filled with ice (Figure 1.2). These valleys had cross sections like a U rather than the V that would have been caused by stream erosion. Waterfalls start from the tops of valley sides and plunge to the rivers in the bottoms of the valleys. This junction between rivers and their tributaries could not have been formed by normal stream erosion (see Rivers). Yosemite canyon in the Sierra Nevada is a U-shaped valley with waterfalls tumbling over its sides and the glacier-carved flat face of Half Dome (Figure 1.3).

Erosion was not restricted to valleys, and broad areas of the flat countryside seemed to have been smoothed off, possibly by flowing ice (Figure 1.4). Places where bare rocks were scraped by debris in the base of the flowing glacier formed "striations" in the direction in which the glacier was flowing. Much of the northern Midwest has disrupted drainage and lakes where it was formerly covered by a thick sheet of ice, and Minnesota claims to be a "Land of 10,000 Lakes."

Geologists also found evidence that the rocks eroded by glaciers at one place were transported great distances and then dropped elsewhere when the glaciers melted. We illustrate this process by looking at the "toe" of a small Alaskan glacier (Figure 1.5). Melting at the lower end of the glacier causes the ice to drop debris that it eroded farther up its valley. This melting forms a ridge of sediment referred to as a "moraine" at the end of the glacier and an "outwash plain" of finer-grained (more easily transported) debris below the moraine.

Some moraines up to 100 miles long show where ice sheets briefly halted as they melted. Blocks of various sizes, from boulders to small houses, are scattered in places where they had no reasonable source. Geologists referred to them as "erratics." Piles of jumbled debris cover the landscape in places where no river could have moved them. Many of these piles have an "aerodynamic" shape, as if they were deposited beneath a glacier and smoothed off by its flow. They are called "drumlins." Some piles of debris show a sinuous pattern. They were deposited by streams flowing beneath glacial ice and are known as "eskers."

In addition to this physical evidence for glaciation on land, geologists also recognized that there had been a rapid rise in sea level as the ice trapped on land melted back into the oceans. Fjords are simply glacial valleys filled by this rising sea level (Figure 1.6). Some irregular

**Figure 1.2** Valley of Rhone River. The glacier filled the entire valley 15,000 years ago.

topography on land was converted into small islands when the land was flooded. The islands of Martha's Vineyard and Nantucket, off the Massachusetts coast, and most of Long Island, New York, are glacial moraines now isolated by sea level rise.

**Figure 1.3** Half Dome, in Yosemite Valley, California. The steep face was carved by a glacier flowing down Yosemite Valley.

**Figure 1.4** Disrupted drainage on glacially scoured landscape in Wisconsin.

**Figure 1.5** Melting toe of Exit Glacier, Alaska. The ridge of sediment at the base of the toe is a moraine, and the sediment in the foreground is an outwash plain.

**Figure 1.6** Fjord in Norway, with glacially carved peaks in the background.

*Peopling of the Americas*

The opening of the Atlantic and Arctic Oceans separated North and South America from the other continents by about 150 Ma (see Pangea in Chapter 3). When the separation was complete, it was impossible for land animals to move between the two Americas and also between them and the rest of the world. Consequently, evolution took very different courses in North America, South America, and in other continents. The separation between North and South America continued until about 3 Ma, when construction of the Isthmus of Panama allowed land animals to move between the two continents.

One of the results of different evolutionary patterns is shown by primates. No primates evolved in North America, and only monkeys developed in South America. Many South American monkeys have prehensile tails (which they use to swing from trees), whereas all African and Asian monkeys can use their tails only for balance.

All primates except South American monkeys evolved in Africa and Asia. They include lemurs in Madagascar, tarsiers in islands of Southeast Asia, baboons in Asia, gorillas and chimpanzees in Africa, and humans in Africa. Humans are presumably related to chimpanzees although humans have 23 chromosomes and chimpanzees and other apes have 24.

Remains of the earliest hominins (humans and their ancestors) have been found in sediments about 6 million years old in Africa. By about 100,000 years ago, humans evolved to *Homo sapiens* (wise man) and began to move around the earth.

By no later than 25,000 BC people had reached Australia, possibly in small boats blown across the straits between Australia and Indonesia. By the same time almost all of Europe and Asia was occupied except for ice-covered areas of northern Europe (see Topographic Effects of Deglaciation). Because northern Asia was not glaciated, people were also living at high latitudes in Siberia.

But there still weren't any people in the Americas. Evidence from numerous archaeological sites shows that people lived in the Americas by about 13,000 BC or slightly earlier. There have been many explanations for the arrival of people in the Americas, but the most popular one is that people walked from Siberia to Alaska. During the coldest part of the ice age, about 15,000 years ago, so much water was held by continental glaciers in Europe and North America that sea level was more than 100 meters lower than it is now, and the Bering Sea was dry land (Figure 1.7). We know the area was used by animals because remains of mammoths have been found in the tiny Pribilof Islands, now isolated in the middle

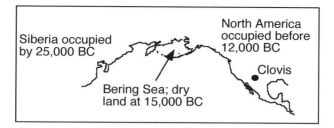

**Figure 1.7** The Bering Sea was land until about 13,000 years ago.

of the Bering Sea. Bands of hunters in the Bering area presumably followed the mammoths and other game, including caribou.

Once people were in Alaska, they could move south by several routes. One route would have been along the North American coast, possibly using small boats to avoid the rugged terrain on land. Another route might have taken people inland and then south along the western edge of the enormous North American ice sheet. By whatever route they used, some of these early people stayed in New Mexico to craft spear points and develop a "Clovis culture" by about 9000 BC. It is the oldest widely recognized culture in North America and is referred to as Clovis for distinctively shaped spearheads found at Clovis, New Mexico.

By 14,000 to 13,000 BC warming of the northern hemisphere had melted enough glacial ice that sea level began to rise rapidly, and the Bering Sea was flooded by 12,000 BC or earlier. This flooding isolated the Americas from the rest of the world and led to the separate development of a Native American society that had no domestic grazing animals, such as horses and cows (see Chapter 4).

Presumably the absence of draft animals prevented Native Americans from inventing wheels. Consequently, they had to transport everything that they needed by carrying it or paddling along rivers in canoes. This lack of the concept of wheels also caused Native Americans to make pottery by coiling ropes of clay around in circles instead of turning the ceramics on a potter's wheel.

The absence of wheels and difficulty of transporting goods may have affected a broad range of technological developments in North America. Native Americans knew how to produce gold and silver from ores because they are relatively easy to smelt. They never learned to smelt base metals such as copper, iron, and lead, however (see Chapter 4).

The absence of iron (steel) and wheels meant that Native Americans were rapidly overwhelmed when Europeans arrived with their weapons, horses, and carts. This disadvantage was greatly increased when enormous numbers of Natives succumbed to smallpox and other diseases that arrived with Europeans who were immune to them because they had already survived epidemics in Europe. These problems were compounded by the belief of many Europeans that Native Americans were simply inferior people who deserved to be exploited. Apparently 90 percent of Native Americans died when Europeans arrived.

### Flooding of the Black Sea

The Bosporus connects the Black Sea with the rest of the world's oceans and is one of the world's busiest waterways. It is also the narrowest

### During glaciation

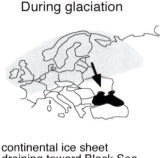

continental ice sheet
draining toward Black Sea

### After glaciation

no continental ice sheet;
rivers draining northward

**Figure 1.8** The Black Sea was full when rivers from an ice sheet (stippled) drained southward. Sea level in the Black Sea was lower after the ice sheet melted and most rivers drained northward.

and most dangerous. Because the Bosporus is the only passage between ports along the Black Sea and the rest of the world, so many ships use it that they are not allowed to enter without a Turkish pilot on board.

The Bosporus hasn't always been open. Since the end of the last Ice Age, there seems to have been at least one time when the Black Sea was completely isolated from the worldwide ocean. We illustrate this process in Figures 1.8 and 1.9. Before about 12,000 years ago (10,000 BC), when a large continental glacier covered northern Europe, glacial meltwater poured down rivers into the Black Sea and kept the water level relatively high.

At the same time as rivers were delivering water to the Black Sea, continental glaciers had extracted so much water from the oceans that global sea level was more than 100 meters below its present level. Thus it is possible that about 10,000 BC the Black Sea was a freshwater lake and may have drained out through the Bosporus toward the Mediterranean.

Figures 1.8 and 1.9 also show what may have happened as continental glaciers melted. The glacier in northern Europe had depressed the surface beneath it, and

AT 10,000 BC

Mediterranean    Bosporus            Black Sea

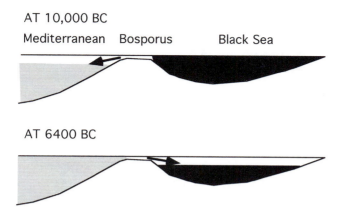

AT 6400 BC

**Figure 1.9** Change in level of the Black Sea in 6400 BC reversed the flow through the Bosporus.

the land was unable to rise ("rebound") as fast as the glaciers melted. Areas that had been covered by the thickest ice were thus left below sea level, forming the Baltic and North Seas. (A similar slowness to rebound also created Hudson's Bay in Canada.)

When the Baltic and North Seas were opened, rivers took water northward and did not drain into the Black Sea until further rebound established the present topography of Eastern Europe. This river diversion caused the Black Sea to shrink, with an ultimate lowering of water level by about 100 meters. Because the northern edge of the Black Sea is a very gently sloping surface, this process would have exposed a band of former sea floor up to 100 kilometers wide. It seems likely that many people settled in this newly exposed, fertile land.

They and their descendents would have prospered until about 6400 BC, when glacial melting raised global sea level to the level of the Bosporus. Some geologists propose that the seawater cascaded abruptly into the basin of the Black Sea, replacing freshwater with typical seawater. This rapid replacement can explain the observation that cores of Black Sea sediments contain freshwater fossils in sediments older than 6400 BC and normal marine fossils in younger sediments just a few centimeters above the freshwater fossils.

The margins of the Black Sea also show evidence of flooding. On the northern margin, rising water apparently moved the shoreline inland at a rate of about 1 meter per day. This rate would have allowed people living in the area plenty of time to move northward, leaving behind buildings that have recently been discovered beneath the Black Sea.

The possibility of a rapid cascade of water through the Bosporus and consequent flooding raises the possibility that the event was the flood described in the Bible. Both the Babylonian epic of Gilgamesh, which is older than the Bible, and the book of Genesis in the Bible describe an enormous flood that affected the entire history of the known world. Flooding in the Black Sea wouldn't have been as dramatic as the flood described in the Bible, but it would have left a record of displacement to become a major event in people's historical record.

But what about the biblical tradition that the flood occurred in Mesopotamia, the area that is now modern Iraq? Naturally there may have been floods there, but could any of them brought Noah's ark to rest near the top of Mount Ararat, as stated in Genesis 8:4: *And the ark rested in the seventh month, on the seventeenth day of the month, upon the mountains of Ararat.* Mount Ararat, which is more than 5,000 meters high? Geologists point out that the ark could not have come to rest there unless sea level was 5,000 meters higher than today, and there simply isn't enough water on the earth to do that. Some people who think the Bible is literally true, however, have always believed Genesis 8:4 and have looked for the ark on Mount Ararat.

### Short-Term (Hundreds of Years) Variations

Wilhelm Gustav Sporer was a 19th-century German astronomer who discovered variations in the intensity of radiation emitted by the sun. The variations are related to changes in the activity of sunspots, and there are numerous "sunspot cycles" of different durations from 7 years to several decades.

Low sunspot activity correlates with low solar irradiance. Low irradiance also leads to low production of "cosmogenic" isotopes such as beryllium-10 ($^{10}$Be). Measurement of these isotopes in tree rings or other datable materials then leads geologists to a history of solar activity in the last few thousand years. We can recognize at least four minima in the last 1,000 years: Oort minimum from about 1000 to 1100; Wolf minimum in the 1300s; Sporer minimum from the middle 1400s to middle 1500s; and Maunder minimum, which lasted for about two centuries and was most intense from middle 1600s to middle 1700s. The Maunder minimum presumably caused most of the Little Ice Age, which ended only about 100 years ago.

Variations in irradiance greatly affected the climate of the last 2,000 years, and we discuss it here.

## Climate of the Last 2,000 Years

The past 2,000 years can be divided into at least five different periods. The difference in average temperatures between warmest and coldest during this time was only about 2°C. Even this small difference was significant, however, and the effects had a major influence on history. The periods graded into each other, and the dates we assign to them are only approximations.

### Roman Optimum (100 BC to 200 AD)

Rome was founded at some time before 500 BC. It remained a local power until it began to expand at about 300 BC. The beginning of the Roman optimum about 100 BC aided the expansion. By 50 BC, Julius Caesar had conquered Gaul (mostly modern France) and had moved into England. He brought Egypt under control within a few years of conquering England.

Other Roman leaders continued to expand the empire. By the early years of the 1st century AD, they had captured Jerusalem and consolidated their hold on much of the Mediterranean. Then they expanded eastward, conquering Turkey, continuing as far east as the Caspian Sea, and almost surrounding the Black Sea.

The Roman Empire reached its maximum extent at about 120 AD. At this time the emperor Hadrian abandoned the idea of subduing Scotland and ordered the building of a wall to separate Roman England from the Scots (Figure 1.10). Further expansion elsewhere was also impossible, and the Romans set about maintaining control of the area they already possessed.

### Vandal Minimum (200 AD to 700 AD)

The end of the Roman optimum gave an opportunity for subjugated people to rebel and for people outside of the empire to invade. Invaders penetrated into the Roman homeland (modern Italy) by 360 AD, and the Visigoths sacked Rome in 410 AD. Attila, the leader of the Huns, stormed into Europe in the middle of the 5th century AD. His armies reached both Rome and Constantinople (modern Istanbul), although he did not conquer the cities. The Huns withdrew eastward after Attila died in 453 AD.

**Figure 1.10** Hadrian's Wall. (*Courtesy of Elizabeth Wolfram*).

*Medieval Warm Period (700 AD to 1350 AD)*

The warmth that developed in the 8th century encouraged Europeans to expand northward. Iceland was first settled in 854, and the first Icelandic parliament was held at Thingvellir in 930 (see Iceland in Chapter 2). England began to produce wine as far north as the Scottish border. Greenland and Newfoundland were settled because Eric the Red and his son killed the wrong people. Eric was thrown out of Norway, then the Orkney Islands, and then Iceland. Fleeing ever westward, Eric established a colony on the southwestern tip of Greenland in the late 10th century AD. Leif, Eric's son, had similar problems and found it necessary to move even farther west. In about 1000 AD, Leif and other Vikings reached North America and established a colony at l'Anse aux Meadows, on the northwestern tip of Newfoundland. The colony was abandoned after only a few years, and the inhabitants moved back to Greenland.

*Little Ice Age (1350 AD to 1850 AD)*

The Little Ice Age encompasses five different episodes separated by brief periods of warmth. Each of them corresponds approximately to a period of low solar intensity. The Wulf minimum of the 14th century was probably responsible for abandonment of Viking settlements in Greenland. The Sporer minimum in the 15th and 16th centuries had a greater effect than the Wulf minimum. It was so cold that the entire Baltic Sea froze solid during the winter of 1422–1423.

The Maunder minimum and the immediately following Dalton minimum kept climates cold from the early 1700s to the late 1800s. The canals of Venice and the Thames at London froze during some years, and there was famine in much of the world. The cold climate in Europe caused food production to decrease and prices to rise. This situation spurred emigration to North America. When people decided to come to America they may not have known that cold weather forced Iroquois tribes to move southward and let people in New York walk across a frozen sea between Manhattan and Staten Island. But this coldness in America really didn't matter to new arrivals. There was so much more land in America that there would be more food per person even in cold years.

High food prices contributed to the French Revolution in the late 18th century. When people complained about the cost of bread, Marie Antoinette, wife of King Louis XVI, is alleged to have said "let them eat cake," which meant the scrapings from cooking pots. She was imprisoned and then executed on October 16, 1793, several months after her husband had been killed.

### Warming since the Middle 19th Century

The protracted Little Ice Age gradually gave way to warmer climates in the late 1800s. We discuss present global warming next.

## PRESENT GLOBAL WARMING

Temperatures are rising in the atmosphere, in the oceans, and on the earth's land surface. What is happening, and can anything be done about it? We discuss the evidence for global warming, the causes, proposed remedies, and possible future consequences.

### Evidence for Global Warming

The evidence for global warming is a mixture of recent observations, historic measurements, and persuasive inferences from less-quantitative information. Figure 1.11 shows the climate change of the past 1,000 years. The middle part of that period is dominated by fluctuations with a trend toward lower temperatures (see Little Ice Age). This cooling ended in the late 1800s, and temperatures began to rise. Several types of

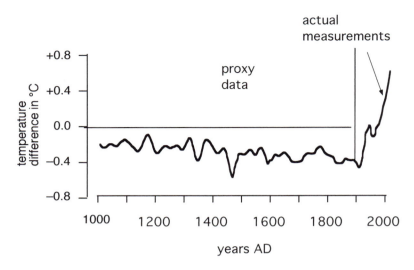

**Figure 1.11** Temperature changes in the northern hemisphere in the past 1,000 years based on data from Intergovernmental Panel on Climate Change (IPCC). Temperatures are shown as deviations above and below an arbitrary 0°C set at average temperatures in the early 1950s. Temperatures since 1900 are based on data from a worldwide network of recording stations. Temperatures before 1900 are synthesized from a variety of proxy data.

evidence show that this normal trend toward increasing temperatures accelerated in the middle of the 1900s:

- In the past 50 years, worldwide atmospheric temperatures have increased about 1°C and ocean temperatures about 0.05°C.
- Both marine and land animals and plants appear farther north in the northern hemisphere than they did 50 years ago.
- Plants in the northern hemisphere bloom one or more weeks earlier in the spring than in the past.
- Both the thickness and extent of ice covering the Arctic Ocean in the summer have decreased dramatically in the past 50 years, and some people predict that by 2050 ships can sail freely across the Arctic during the summers.
- The permanently frozen ground (permafrost layer) has melted in some places in the far north, causing collapse of buildings whose foundations are rooted in it.

The southern margin of the Sahara desert has steadily encroached southward into the northern part of the water-rich area of central Africa (the Sahel). The southward movement of the desert is particularly noticeable in Mali (Figure 1.12). The Niger River in northern Mali provides water for an agricultural economy in semiarid lands that end abruptly at the southern end of the Sahara desert. The farmers and desert nomads lived with little conflict throughout much of their history, but that peacefulness began to change in the 20th century. As the Sahara desert became even drier and

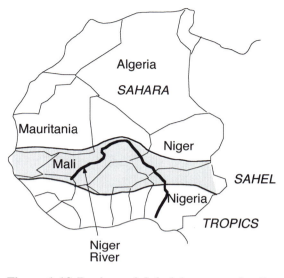

**Figure 1.12** Region of Sahel between the Sahara Desert and tropical forests.

oases dwindled, nomads moved southward into agricultural land. In the 21st century, Mali is now enduring a fully developed civil war.

**Causes of Global Warming**

The warming of the last half-century has been accompanied by rapid increases in the abundances of greenhouse gases. Within the last century, data on the composition of the atmosphere are based on actual measurements of air samples, but earlier information comes largely from air samples in bubbles in ice cores (see Long-Term (Tens of Thousands of Years) Cycles). Figure 1.13 shows changes in the abundances of $CO_2$ and $CH_4$ in the atmosphere in the past 1,000 years.

The increases in both $CO_2$ and $CH_4$ are clearly caused by human activity. Cows, horses, sheep, and other "ruminants" emit methane as a waste product, and the increase in atmospheric methane that began perhaps around 1750 is almost certainly a result of the increase in the number of grazing animals that people raised for food and transportation.

The increase in the abundance of carbon dioxide that began in the late 1700s occurred at the start of a period commonly referred to as the "industrial revolution." It was initially fueled by coal, which was used for heating in fireplaces and stoves and also to run steam engines for a large number of industries. Oil and natural gas became important fuels

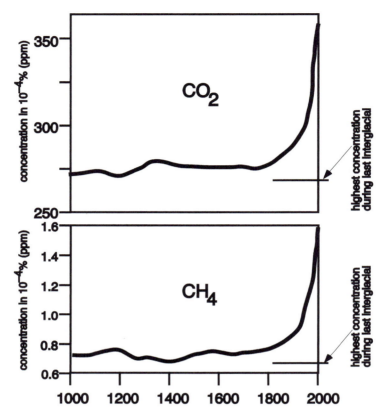

**Figure 1.13** Variations in the concentrations of $CO_2$ and $CH_4$ in the last 1,000 years. The highest values reached during the previous interglacial period are based on measurements in ice cores.

in the 1800s, particularly after the discovery of oil in the United States (see Chapter 4). The development of cars and trucks accelerated the use of oil, especially in developed countries such as the United States.

The increases in concentrations of greenhouse gases are clearly responsible for some of the temperature increase. Higher solar intensity, however, is probably responsible for some of it. Climate scientists have not been able to determine what proportion of the temperature increase is attributable to these or other causes.

**Reducing the Rate of Global Warming**

Most climate scientists understand global temperatures will continue to rise even if we stopped all industrial activity immediately (which is obviously impossible). The atmosphere contains so much $CO_2$ and other greenhouse gases that they cannot be brought down to preindustrial levels for many decades, possibly the rest of the 21st century. The realization that global warming will occur regardless of people's efforts to stop it means that we have to plan for the change.

The fact that global warming is inevitable for much of the 21st century doesn't mean that we shouldn't look for ways to reduce emissions of greenhouse gases. An effective policy of reducing emissions might prevent global warming from continuing for many centuries.

There is little that we can do about $CH_4$, but we can take two approaches to reducing emissions of $CO_2$. One is to consume energy at the same rate as now but to switch to fuels other than coal, oil, and gas. "Biofuels" are synthesized from plants and can replace oil in all its uses, particularly cars and trucks (see Chapter 4). Biofuels emit $CO_2$ when burned, but they do not increase the $CO_2$ content of the atmosphere because it is the same amount that the plants would have produced if they were eaten by animals, burned in fireplaces, or simply left to decay on the ground.

Other alternative sources of energy produce electricity without any emission of $CO_2$ (see Chapter 4). They include solar cells, wind power, water power, and nuclear energy. Any effort to use these sources of electricity to run automobiles, however, requires the development of better batteries than are currently available.

The most controversial source of energy that doesn't produce $CO_2$ is nuclear power. The earth contains abundant supplies of nuclear material, and the amount of electricity that can be produced is theoretically limitless. Unfortunately, nuclear reactors also produce radioactive waste products that must be stored in a safe place for up to 250,000 years (see Chapter 4). At present, society hasn't decided whether the danger from nuclear waste is greater or less than the benefit of using nuclear power to generate energy.

A second approach to decreasing $CO_2$ emissions is simply to reduce our use of energy. The possibilities are endless, and we can mention only a few of the more obvious ones. Instead of driving individual automobiles, we could use public transportation or car pools. If we do use individual automobiles, we can require them to be more fuel-efficient (yield more miles per gallon). We can keep houses and public buildings a few degrees cooler in the winter and warmer in the summer.

The problem with alternative energy sources and energy conservation is that they are slightly more expensive than fossil fuels at present and also somewhat inconvenient. Most people want to drive their own cars and keep their houses at the same temperature during both winter and summer. People do not want to pay more for energy than they pay for coal, oil, and gas.

### Consequences of Global Warming

Most consequences of global warming will have a negative impact on society, but some may be positive.

Negative impacts include the shift in climate zones that will accelerate the changes in the Arctic and Sahara that we discussed earlier. Rising sea level will flood coastal areas and may make countries such as the Maldives uninhabitable (see tsunami in Chapter 2).

Another surprising negative consequence of global warming may be extreme cooling of North America and Europe. The present North Atlantic current brings warm tropical water past the east coast of North America and around northern Europe (see Ocean Currents). This current exists only because water farther north in the Atlantic Ocean is so cold and salty that it has a higher density than normal seawater. This dense water sinks to the floor of the North Atlantic Ocean and then passes along the floors of the Indian and Pacific Oceans before it returns to the North Atlantic near the surface.

Sinking of the North Atlantic water will stop if more of the Greenland icecap melts because the fresh water from the ice will dilute the seawater and make the density so low that the water cannot sink. Then if the water can't sink, the North Atlantic current will shut down, and North America and northern Europe will become much colder.

The North Atlantic current has stopped twice since the last ice age; once at about 10,000 BC and once at about 6,000 BC. Both stoppages required only 10 to 20 years to convert the North Atlantic Ocean from normal circulation to one without a warm current. The shut off of the current not only brought cold weather to North America and Europe but also aridity farther south.

Some consequences of global warming, however, may be beneficial. One is the movement of agricultural activity farther north, and farming has already begun in southern Greenland. Another benefit is the possibility of increased shipping across the Arctic Ocean.

An airplane flying at slightly more than 500 miles (800 km) per hour requires about 10 hours to fly the 5,000 miles (8,000 km) between London and Tokyo. Most modern cargo ships sailing at about 30 miles (50 km) per hour need 4 to 5 weeks to cover the approximately 15,000 miles (25,000 km) between London and Tokyo via the Atlantic Ocean, Mediterranean Sea, Suez canal (see Suez canal), Red Sea,

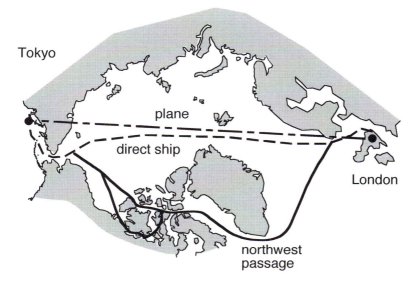

**Figure 1.14** Potential shipping routes across the Arctic Ocean as warming continues.

Indian Ocean, and then north to Tokyo through the western Pacific Ocean.

Global warming may make the shipping route much shorter and faster by the middle of the 21st century Figure 1.14). Scientists working for the Intergovernmental Panel on Climate Change (IPCC) have noted such a rapid disappearance of ice in the Arctic Ocean during the summer that they have predicted that the Arctic Ocean could be ice-free and open to shipping during summer months by 2050 or earlier. At that time ships could cross the Arctic Ocean from Europe to Asia in less than 2 weeks.

We may not have to wait until the Arctic Ocean is ice-free to have a fast route for ships between Europe and Asia. Gradual warming has already left one or more routes through islands north of Canada passable during the summer months. This route cuts 2 to 3 weeks off the shipping time on the present route through the Suez Canal and the Indian Ocean.

We emphasize that all of the potential climate changes discussed here have been rapid, with most of the change completed within a few tens of years.

**CLIMATIC EFFECTS OF VOLCANOES**

Volcanoes that put enormous amounts of ash into the atmosphere can affect climate for some number of years that depend on the size

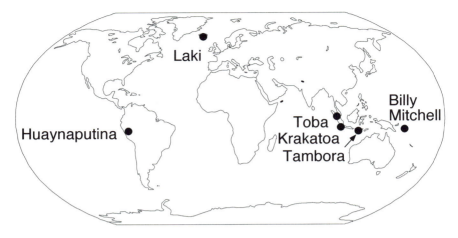

**Figure 1.15** Locations of volcanoes discussed in the text.

of the eruption (see Chapter 2). We discuss the effects of three volcanoes in Indonesia (Toba at about 75,000 BC, Krakatoa in 535 AD, and Tambora in 1815); an eruption at Laki, Iceland, in 1783; and eruptions of Mount Billy Mitchell in the Solomon Islands in the late 1580s; and Huaynaputina in 1600 in Peru (Figure 1.15).

**Toba**

All people, wherever they live, form a single species of men and women who can breed with each other. Despite this biological identity, however, people can be recognized by an astounding variety of "racial" characters. Skin colors include white, black, brown, red, and yellow. Hair and eye colors are more variable than skin colors. Eyes are round or oval and slanted, with all shapes in-between. The diversity increases if we look at other parts of our bodies.

So, how did one interbreeding population become so diverse? Many anthropologists believe that this could not have happened if our ancestors developed in Africa, began to move out about a hundred thousand years ago, and gradually adapted to new environments and new climates. Paleontologists studying evolution of fossil groups know that gradual change leads to small differences. Consequently, anthropologists began to search for some drastic occurrence that isolated small groups of people in different places on the earth and caused inbreeding of groups so small that a few characteristics came to dominate each group. Some anthropologists believe that the world's population may have dwindled to no more than 10,000 people. The only possible catastrophe seems to have been a large volcano.

The largest eruption that anyone can find in the last 100,000 years was at the present site of Lake Toba. The lake is in the mountainous center of Sumatra, has an elevation of 400 meters, a length of 100 kilometers, and a width of 30 kilometers. The lake apparently is the site of a volcano that completely blew apart about 75,000 years ago.

The age of the eruption is known because rocks formed by it have been dated. The size of the eruption can be estimated by measuring the size of the lake, which was formerly a mountain, and measuring the amount of debris in Indonesia and surrounding seas. This information makes it clear that Toba erupted with a violence that put so much ash in the atmosphere for such a long time that it caused six of the coldest years in the last 100,000 years and accelerated the last glacial period (see Vostok discussed earlier).

All of these effects show that the Toba eruption may have caused the decimation of the population that led to the diversification of people into so many different races.

### Krakatoa, 535 AD

Two thousand years ago, the islands of Sumatra and Java were joined as one island. The explosion of Krakatoa in 535 AD created the Sunda Strait that now separates the islands. The eruption also had enormous climatic effects, particularly because it occurred during the cold period that separated the Roman optimum from the medieval warm period (see Short-term (hundreds of years) Variations discussed earlier).

According to British archaeologist David Keys, the explosion unleashed a catastrophe that affected the entire world. One of the most serious consequences was destruction of predators in central Africa around the area of present Uganda. These predators had maintained control of the local population of a type of rat that carries the bacteria that cause bubonic plague (Black Death). The plague is transmitted to people by fleas that bite infected rats, and it spreads quickly once rats arrive in an area. The plague arrived in the eastern Mediterranean in 541 AD, only 6 years after the eruption of Krakatoa. Nearly half of the population of the area died quickly.

In addition to the plague, the eruption of Krakatoa may have sent another horror to Europe. People who herded cows in northeastern Asia suddenly found that there wasn't enough vegetation for cows to live on. Horses can live on forage that cows cannot digest, however, and societies based on horses began to flourish. They displaced people with a cattle-based economy westward into Europe. The Huns, led by Attila, followed westward in the 5th century, and the mix of different people in Europe had changed permanently by the time the Huns left.

### Tambora

Tambora erupted in 1815. It put so much ash in the atmosphere that 1816 was regarded as the "year without a summer," or "1800 and froze to death." The temperatures were unusually low in eastern Canada, the northern United States, and northern Europe. Frost in May and snow in June as far south as Pennsylvania killed most of the crops that had been planted in the spring. Grain prices more than tripled in eastern North America, and people who didn't die from the cold were unable to buy food.

### Laki, Iceland

The Laki rift zone is part of the Great Icelandic rift (see Chapter 2). It erupted enormous volumes of basalt lava and ash for 8 months in 1783. The ash was accompanied by 80 tons of sulfur dioxide aerosols, which caused widespread acid rain. The combination of ash and acid rain destroyed so many crops and animals in Iceland that one-fourth of the population died from the famine.

The cloud of acid rain also spread over Western Europe in the summer of 1783. It killed crops and poisoned animals, particularly in Britain and Ireland. As in other famines in Europe, more people were forced to immigrate to the United States.

### Mount Billy Mitchell, Solomon Islands

Growing trees add one "ring" of wood each year, with the youngest on the outside and the oldest in the middle. Wide rings are formed by rapid growth during years that are warm and rainy, but narrow rings develop because of slow growth during years of cold and drought. These differences in widths of rings provide a history of weather during the lifetime of the trees.

Trees that have long lives, such as the bald cypress of coastal Carolinas and Virginia, can record several hundred years of history. A study of this history reveals two periods of drought and summer cold in the late 1500s and early 1600s AD.

One was from 1587 to 1589. An English colony had been established on Roanoke Island, North Carolina, in 1585, but when ships arrived in 1590 to resupply the colony, the crew found no people and

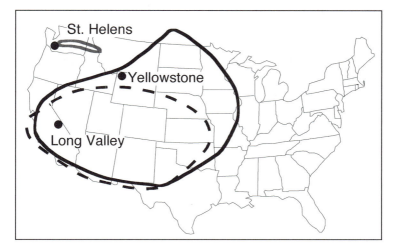

**Figure 1.16** Distributions of ash from the most recent eruptions at Yellowstone and Long Valley. St. Helens' ash is for scale.

only a hand-lettered sign with the word CROATOAN. The sailors assumed that survivors had gone to the mainland and blended in with the Native Americans living in the area that is now the "Croatan" forest. Because no one knows the fate of the colonists, we now refer to the place where they attempted to settle as the Lost Colony. A principal reason for the drought may have been massive eruptions of Mount Billy Mitchell on Bougainville, Solomon Islands, in 1585 and 1587.

### Huaynaputina, Peru

A drought from 1606 to 1612 shortly followed the eruption of Huaynaputina, Peru, in 1600 AD. The drought occurred when the English were trying to establish a colony at Jamestown, Virginia. Death from malnutrition was high in the colony, with only 38 of 104 colonists surviving the first year (1607) and more than half of the new arrivals dying in subsequent years.

### Possible Future Eruptions of Yellowstone and Long Valley

Yellowstone is a caldera with a diameter of about 50 km (see Chapter 2). Its last eruption was 664,000 years ago, and the ash covered most of the western United States (Figure 1.16). A smaller volcanic area at Long Valley, California, erupted most recently 760,000 years ago and covered a smaller area with ash. An eruption of either of these volcanoes

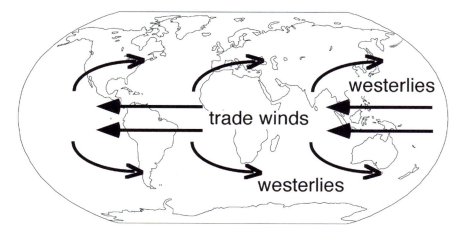

**Figure 1.17** Global wind patterns.

in the near future would devastate large areas with ash fall and might even make major changes in climate.

### WIND

Winds are caused mostly because the earth rotates from west to east at a speed of approximately 1,500 kilometers per hour (kph) at the equator. The earth carries its atmosphere along as it rotates, but there is enough slippage that the air at the equator does not move quite as fast as the earth beneath it. This difference causes wind in the tropics to blow steadily from east to west, creating "easterlies" (trade winds) that blow at average speeds of 15 to 30 kilometers per hour (Figure 1.17).

Another type of air movement is caused when sunlight at the equator heats the air to higher temperatures than to the north and south. This hot air rises, cools off, and then descends to the earth's surface at latitudes of about 20°N and 20°S. Because the cooling causes the air to lose moisture by raining, the descending air is dry and creates most of the world's deserts slightly north or slightly south of the equatorial tropics.

Some of the descending air returns to the equator, but much of it moves northward in the northern hemisphere and southward in the southern hemisphere. As the air enters temperate regions, it still has the west-to-east speed it had at the equator. Because the speed of rotation of the earth decreases from the equator to the poles (where it is zero), air in temperate latitudes is moving faster toward the east than the earth beneath it. This difference results in the "Coriolis effect" and causes prevailing winds, called "westerlies," to blow from west to east at highly variable speeds.

The Coriolis effect is also important in the development of hurricanes and typhoons. When a small area of hot air develops near the earth's surface, it rises and thereby creates an area of low pressure. The hurricanes that reach North America are formed from cells of hot air that move westward from the Sahara desert in Africa. Air that moves toward these cells of low pressure begins to follow the Coriolis effect and rotate counterclockwise around the "eyes." (Storms rotate clockwise in the southern hemisphere.) The strength of a hurricane depends on the amount of heat it gains as it moves across the Atlantic Ocean.

Hurricanes are graded on the Saffir-Simpson "intensity" scale, which was originally developed in 1969 by engineer Herbert Simpson and Robert Saffir, who was the Director of U.S. National Hurricane Center. The scale grades hurricanes according to their highest wind speeds as follows: category 1, speeds of 74–95 mph (119–153 kph); category 2, 96–111 mph 154–177 kph); category 3, 111–130 mph (178–209 kph); category 4, 131–155 mph (210–249 kph); category 5, >155 mph (249 kph). Hurricane Katrina was "only" a category 4 when it hit the U.S. Gulf coast in August 2005, and we describe it next.

### Hurricane Katrina

The delta of the Mississippi River continually advances into the Gulf of Mexico by deposition of sediment eroded from the middle of the United States. As the river reaches the low elevation near the coast, it breaks into "distributaries" that follow different paths to the ocean. These distributaries make a complex pattern that gives the name "birdfoot" to the delta. The borders of the distributaries are "natural levees" with tops that are above river level, but the "interdistributary areas" are swamps that are below the rivers (further explanation in Rivers).

French explorers began charting the Mississippi River in the late 1600s. In 1718 French ships sailed 160 kilometers up the Mississippi from its mouth and found a place where the natural levees were high. They established a city on this high ground and named it for the Regent of France, Philippe II, the Duc d'Orleans. New Orleans became an American city in 1803 as part of the Louisiana Purchase (see Lewis and Clark).

When New Orleans expanded beyond its original "French Quarter," there was no place to go except into the swamp away from the natural levees. By 2005, almost the entire city of nearly 500,000 people was several meters below the level of the Mississippi River. New Orleans was kept dry and habitable only by a complex series of dikes to keep water out and an equally intricate set of pumps that removed water that seeped in.

Because of its location, New Orleans was a disaster waiting to happen. The catastrophe arrived on August 25, 2005, with Hurricane Katrina, whose wind and high water broke some of the dikes and stopped the pumps when the electrical system was destroyed. The eye of Katrina moved along the eastern edge of the delta of the Mississippi River and came ashore in the general area between New Orleans and Waveland, Mississippi. Thus the city suffered not only wind damage but also flooding. The French Quarter and other areas along the natural levee were above the level of the Mississippi River and remained dry. Areas of the city below the river level, however, soon filled with water, which covered the streets and reached the upper stories of houses in areas that were particularly low.

Many of the residents of New Orleans had fled the city after the mayor ordered an evacuation on August 28, but about 60,000 people were either unable or unwilling to get away. The people who were trapped in New Orleans faced many days of tenuous survival. Some were rescued from flooded houses and taken to the partially flooded Superdome, where supplies of food and water were inadequate. Some people scavenged stores for food, and some reached the Convention Center, where fights broke out over scarce resources. About 2,000 people died, although the exact figure was unknown one year later because several hundred were listed as "missing."

Although most of the news about Hurricane Katrina was directed toward the city of New Orleans, other parts of the Gulf Coast were also devastated. The counterclockwise movement of the hurricane brought the highest winds against the shoreline of Mississippi just east of New Orleans. In addition to causing direct damage, the wind piled up a "storm surge" of seawater more than 10 meters high that rushed ashore and obliterated houses and other structures up to several kilometers inland. Because people who wanted to escape could easily drive inland, only about 200 people in Mississippi were killed by the hurricane, but the total damage was several hundred billion dollars.

Two years after the hurricane, the future of New Orleans and the Mississippi Gulf coast was still uncertain. The dikes around New Orleans had been rebuilt and the water drained out, but more than half of the pre-hurricane population had permanently moved elsewhere. The major questions were how much of New Orleans should be rebuilt, and who should pay for it. New Orleans is clearly invaluable as a port on

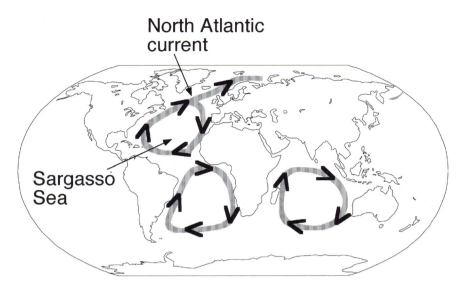

**Figure 1.18** Gyres (currents) in the Atlantic and Indian Oceans.

the Mississippi River, and it is a significant part of America's cultural heritage, but people wondered if the entire city should be rebuilt. Many people thought that rebuilding the city would waste money on efforts that could easily be destroyed by another disaster. The problem was a national one because federal money would have to be spent to maintain the dikes and pumps even if all of the buildings in New Orleans were financed privately.

Federal money wouldn't have to be spent directly on rebuilding along the Mississippi coast, but it might be required indirectly. Commercial insurance companies will not offer insurance against the risk of rising water, which includes floods and also storm surges such as the one caused by Katrina. Consequently any person who wants to build along the coast will have to consider the risks of further damage and whether it will be possible to get insurance to cover them.

**OCEAN CURRENTS**

Ocean currents are established by roughly the same forces that cause winds. At the equator, trade winds blowing toward the west link with surface water to form currents that flow predominantly westward. Where these currents are diverted to the north in the northern hemisphere or the south in the southern hemisphere, they begin to flow back to the east. This circulation pattern ("gyres") occurs in all of the world's oceans (Figures 1.18 and 1.19).

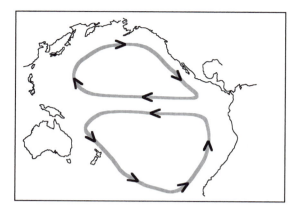

**Figure 1.19** Gyres (currents) in the Pacific Ocean).

Currents in the North Atlantic have been extremely important in history. One current brings warm water from the south to the ocean west of northern Europe, and the prevailing westerly winds keep northern Europe warmer than land at similar latitudes in Canada. A second historical issue is the importance of avoiding the central part of the ocean, particularly during the days of sailing ships. This area was known as the "Sargasso Sea" because much of its surface is covered by floating mats of the seaweed (algae) *Sargassum*. This seaweed occupies an area of the ocean where the absence of strong currents and strong winds caused sailing ships to move very slowly or, sometimes, not at all. Some seafarers referred to the area as the "doldrums."

The problems of the Sargasso Sea led to many myths: sea monsters that devoured ships lived in it; ships that entered it never came out. The tales were even more bizarre than modern ones about the "Bermuda Triangle," a fictional area that overlaps the western edge of the Sargasso Sea.

The pattern of currents in the Pacific Ocean was also extremely useful to mariners in the days of sailing ships. One example is the route of the "Manila Galleon." In the early 1500s, the Spanish became the first Europeans to dominate the Pacific Ocean. They came to the Pacific from both sides, occupying the west coast of North and South America and establishing trading ports in the Philippines and other islands in the southwestern Pacific. Then they looked for the best routes for ships to follow as they carried trade goods both eastward and westward across the Pacific.

The principal route was between Manila, in the Philippines, and Acapulco, Mexico (Figure 1.20). The problem was to use wind and current patterns to find the best paths across the Pacific in each direction. Fortunately, the Spanish mariners already knew about the necessity of finding apparently circuitous routes across oceans because of their experience in the Atlantic. They soon discovered that the Pacific Ocean contained currents and wind patterns similar to those in the North Atlantic. They found that the quickest route from Manila to Acapulco was through the northern part of the ocean in order to take advantage of westerly winds and eastward-flowing currents. On the route back from North America

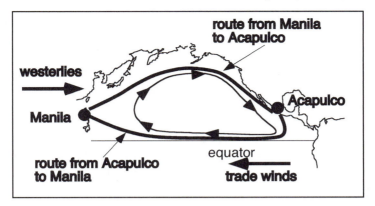

**Figure 1.20** Route of the Manila Galleon.

to the Philippines, however, the fastest route was near the equator, where both winds and currents flowed westward.

Following these routes, the Manila Galleon (or several ships together) made one or two round trips each year between 1565 and 1821, when Mexico won its independence from Spain. *To* Mexico the galleon brought spices from islands in present Indonesia, silk cloth from China, and artifacts from all over the western Pacific. *From* Mexico the galleon brought mostly silver for China and other trading partners.

Airplane travel across the Pacific today is clearly not affected by ocean currents. It is, however, dependent on the westerly winds. Philippine Airlines uses favorable westerlies to fly nonstop from Manila to Los Angeles. On the trip from Los Angeles to Manila, however, westerly headwinds make it necessary for flights to make a refueling stop in Guam.

**El Nino**

The normal pattern of temperatures in the surface water in the equatorial Pacific Ocean shows an increase of a few degrees from the Peruvian coast westward to Australia and other lands in the western Pacific. At irregular times ranging from 2 to 10 years, however, the temperature reverses so that surface temperatures become warmer from the west toward the coast of South America. The arrival of warm water near the Peruvian coast usually comes near the end of the year, around Christmas. Consequently, the reversed weather pattern is referred to as El Nino Southern Oscillation (ENSO), based on "El Nino," the Spanish term for the Christ Child.

Most El Ninos last 1 or 2 years, but they have serious effects even during that short period of time. Some of the increased rainfall in the coastal region is a welcome change to the normal aridity, but too much rain can cause flooding. High rainfall along the coast also prevents water vapor from being blown eastward to the Andes and results in drought in the mountains. Probably the most serious effect of El Ninos, however, is in the ocean just offshore, where the warm surface water is too light to sink. This inability to sink means that nutrient-rich deep water is not forced to the surface and nourish the fish that are the basis of the very profitable Peruvian fishing industry.

The historical record shows one time when an El Nino seems to have lasted for several decades. It apparently destroyed the Moche civilization, which flourished for almost 1,000 years on the coast of Peru near the present city of Trujillo.

The fertility of the valleys near Trujillo attracted people more than 1,000 years before the Spanish arrived in the early 1500s. The most important of the pre-Columbian civilizations was the Moche culture, known for its ceramics and other artwork. The Moche developed an agriculture based on an extensive series of dikes and canals that brought fresh water to crops in fields blessed with year-around sunshine. In addition to irrigation systems, the Moche also built pyramids and temples of adobe (mud).

The valleys near Trujillo now, however, contain only remnants of the ancient irrigation systems and other structures. The pyramids and temples are partly washed away, with gullies down the sides, and the mud bricks are rounded and displaced. The irrigation canals and dams have holes that make them useless. The former sites of numerous buildings can be located, but the buildings themselves are mostly gone.

Radiometric dating, partly of human remains, shows that the destruction of the Moche civilization occurred during a few tens of years in the middle of the 6th century AD. The ruins of Moche sites suggest that the disaster was initiated by severe rain and flooding that washed away the mud blocks that the Moche used for construction and probably also destroyed much of the agricultural land.

But why would there have been heavy rain and floods in a region that is generally arid? The Trujillo region now receives an average of less than 1 inch of rain per year and depends on rivers coming from the Andes to deliver all of the water for the people who live there and for their agricultural production. Presumably some special climatic event was responsible for the deluge that destroyed the Moche.

Although a normal El Nino couldn't have the drastic effects that led to the demise of the Moche, an extreme (mega) El Nino might have been responsible. A good test of that possibility is provided by climate data recorded in ice cores from the Quelccaya ice cap in the Andes Mountains of southern Peru.

The Quelccaya cores have been dated by counting annual layers and also by the presence of ash from the 1600 AD eruption of Huaynaputina. Very thin accumulations of ice occurred from 563 AD to 594 AD. This 30-year period of drought in the Andes was presumably a time of heavy rain and floods in the coastal area occupied by the Moche, and coincide with the beginning of destruction of the Moche civilization. Consequently, these data confirm that the end of the Moche culture was caused by heavy rain during an extreme and long-lived El Nino.

## RIVERS

One of the major problems faced by early geologists was why rivers flow in valleys. This sounds strange because everybody knows that water flows to the lowest level it can reach. The question, however, is which came first—the rivers or the valleys?

We can simplify the problem with this, admittedly absurd, view of how valleys might develop before rivers. Imagine that valleys were cut by groups of enormous prehistoric bulldozers driven by people who had no idea what they were doing. In this situation, there is no reason why rivers and tributaries should flow into each other smoothly ("at grade level"), and we would expect waterfalls wherever two valleys intersect each other (perhaps also wrecked bulldozers).

The problem was solved by John Playfair, a Scottish geologist who lived in the 19th century. He realized that rivers and their valleys generally join at grade level, and constructed a much simpler explanation for the evolution of river systems. Tributaries that flow into rivers at grade level presumably started their valleys at the river and lengthened themselves by eroding backward from the river. This process is called "headward erosion," and it is the process by which all river systems become longer and more complex through time (Figure 1.21).

A river and all of its tributaries form a "drainage basin" that is separated from other basins by a "drainage divide" (see Missouri River). Divides shift continually as river systems with higher gradients "capture" land and parts of rivers from systems with lower gradients (see Atchafalaya River).

Rivers undergo many changes from their headwaters to their mouths. Their gradients are steep near their heads and decrease downstream until the rivers are nearly flat where they flow into their ultimate destinations in the ocean, lakes, or basins of "interior drainage." (see Route Across the Great Basin). Where gradients are high, rivers erode mostly downward and cut steep-sided valleys. Farther downstream, rivers "meander" from side to side and cut valleys that are wider than the rivers

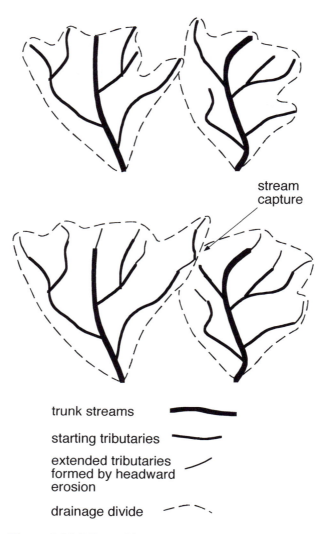

stream
capture

trunk streams

starting tributaries

extended tributaries
formed by headward
erosion

drainage divide

**Figure 1.21** Effect of headward erosion on enlarging
drainage basins. One stream in the drainage basin to
the left encroached into the basin on the right and
captured one of its tributaries.

themselves. These valley floors commonly become "floodplains" when
rivers flood over their banks and cover the entire valley.

As meandering rivers continue to erode, some of them leave parts
of their former floodplains high on the sides of the valleys. These flat
areas are "terraces," with older ones higher on the valley walls. The age

differences of the terraces allow archaeologists to date changes in the civilizations that occupied a valley over a long period of time.

Deposition and erosion both occur in meandering rivers. Erosion is dominant on the outside of meander bends, causing the meanders to become larger as deposition of "point bars" occurs on the inside of the bends (Figure 1.22). Point bars are important sources of gold and other heavy minerals that are dropped by rivers, and we describe one of them in our discussion of the gold industry in Chapter 4.

Deposition becomes increasingly important as rivers move downstream. Some of them deposit bars within their channels, causing rivers to flow around them in complex patterns. One of these areas has made the position of the U.S.-Mexico border uncertain along part of the Rio Grande River (see Chamizal section).

Rivers become mainly depositional when they come close to their mouths. Most of them deposit so much sediment that their single channels break up into smaller ones. These individual channels branch apart downstream and are referred to as "distributaries." Water that flows over the banks of these channels during floods deposit sand and silt in "natural levees" that rise above normal river levels. Rivers that form deltas in the ocean or lakes commonly consist of numerous distributaries surrounded by natural levees. This pattern had a major effect on the disaster caused by Hurricane Katrina.

The amount of water flowing down rivers fluctuates greatly during the year. Most rivers are in flood for about 1 month of the year, and almost 90 percent of the water carried by the river during the year flows during this month. The annual flooding of the Nile River has affected the lives of people in the valley for more than 5,000 years, and we discuss it in the section on the Nile River.

The water in almost all rivers ultimately reaches the oceans. A few rivers, however, flow into basins of "interior drainage" and soak into the ground at their mouths. We discuss them in the section on the Great Basin and also in Armageddon in Chapter 2.

### The Missouri River and the Lewis and Clark Expedition

Napoleon Bonaparte needed money in 1803. Lots of it. So he looked around for something to sell and found that the American government wanted New Orleans because control of the city would give the burgeoning population along the Mississippi River complete access to the ocean. Thomas Jefferson's government offered to buy New Orleans, but the French wanted to sell more—the whole Louisiana Territory. For a final price of some $15 million, American negotiators in Paris bought the territory without bothering to inform Congress or the American public.

**Figure 1.22** Point bar in American River, California.

Now the question became "What had America bought?" The Louisiana Territory clearly contained the Mississippi and Missouri Rivers and extended as far west as the Rocky Mountains. It was large enough to double the size of the United States, but no one knew where the borders were or how far it extended.

Jefferson's next move was to send a party out to explore the region and also, hopefully, to find an easy portage between the headwaters of

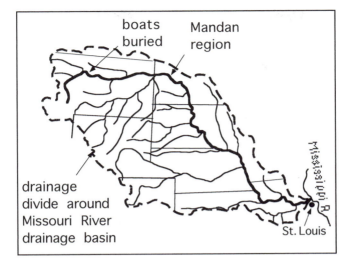

**Figure 1.23** Drainage basin of the Missouri River (heavy line). The drainage divide (dashed line) surrounds the headwaters of all of the tributaries that drain into the Missouri River.

the Missouri and the headwaters of a river draining into the Pacific. For leaders of the expedition he chose two army officers—Meriwether Lewis and William Clark. On May 14, 1804, the party sailed into the mouth of the Missouri River (Figure 1.23). They had to row upstream during the spring flood, and by winter they were near the center of North Dakota, where a few thousand Mandan lived.

Lewis and Clark settled there, and during the winter a fur trader named Pierre Charbonneau wandered into the camp with his wife, who was about to give birth. He wanted Lewis and Clark to hire him to accompany the expedition, but they weren't interested until they learned that his wife had been stolen from the Shoshones. So they hired him in order to have his wife as an interpreter when the party reached Shoshone territory farther west. Her name was Sacajawea.

In the spring the party rowed onward until the river became impassable, and they buried their boats and proceeded onward by land. Sacajawea persuaded the Shoshone chiefs to help the exploring party across the mountains, which formed the drainage divide between the Missouri basin and the rivers that drained westward.

The mountains (now mostly in Idaho) were much higher than Lewis and Clark anticipated, and they barely made it across before winter set in. On the western side, the party encountered the Nez Perce, who were very helpful and directed the party toward tributaries that flowed into the Columbia River. The explorers hurtled down the river in small canoes and had to walk along the banks in the most treacherous places.

On November 24, the expedition came close to the Pacific Ocean and had to build another winter quarters. They put their buildings on what is now the Oregon side of the river and called it Fort Clatsop after the local tribe. The party made it through a winter of almost incessant cold rain and remained in camp until March 23, 1806.

Then it was back across the mountains by an easier route than they had taken on the way west. They took various routes until they found their buried boats and set off down the Missouri. When they reached the Mandan area, Pierre Charbonneau asked for $500 as payment for his services. When he had his money, he took his wife and son, and he and Sacajawea disappeared from recorded history.

Moving rapidly down the Missouri, Lewis and Clark reached St. Louis on September 23. They had succeeded in their principal mission—to map the Louisiana Territory. Because of the immensity of the mountains, however, they failed to find an easy portage between the headwaters of the Missouri River and the headwaters of the Columbia River. Numerous efforts were made to find ocean routes around North America (see Panama Canal section), but travel across the United States would remain difficult until the completion of the transcontinental railroad in 1869.

### Atchafalaya and Mississippi Rivers

The Atchafalaya River is a distributary that branches away from the Mississippi River about 100 miles north of the Louisiana shoreline (Figure 1.24). From this point, the Atchafalaya travels almost directly south through the swamplands that are the principal location of the "Cajun Country" of Louisiana. The Atchafalaya's route to the Gulf of Mexico is much shorter than the route of the Mississippi River. The Mississippi meanders erratically southward and does not enter the Gulf of Mexico until it reaches the end of the very long Mississippi delta (see Hurricane Katrina section).

**Figure 1.24** Atchafalaya and Mississippi Rivers in southern Louisiana.

The difference in routes gives the Atchafalaya a gradient almost twice as steep as that of the Mississippi. If the join between the two rivers were left in a natural state, this difference in gradients would cause the Atchafalaya to "capture" almost all of the water in the Mississippi and become the principal outlet for water in the Mississippi drainage basin. Capture of the Mississippi water would have two important consequences. One result would be flooding of the area around the Atchafalaya, drowning towns and homesteads (and possibly flushing a lot of alligators into the Gulf of Mexico). The second result would be the complete elimination of New Orleans as a seaport. The Mississippi River would dry up so completely that all shipping from the American interior would have to be rerouted down the Atchafalaya.

In 1954 the U.S. government decided that these consequences were too disastrous to be permitted. The government authorized the U.S. Corps of Engineers to construct barricades that would control the flow of water into the Atchafalaya and Mississippi Rivers. At present, the Corps allows about one-third of the water entering the barricades to flow into the Atchafalaya and two-thirds into the Mississippi.

### Chamizal and the U.S.-Mexico Border

Much of the original border between the United States and Mexico followed the Rio Grande River between the United States and Texas. This arrangement encouraged the rapid growth of both El Paso, in the United States, and Juarez, across the river in Mexico. At this time, the river followed a very winding course between the two towns, but then it began to shift toward the south (Figure 1.25).

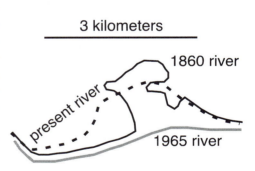

**Figure 1.25** Changes in route of the Rio Grande River between El Paso and Juarez.

By 1960, the river had straightened its course and moved more than 2 kilometers south from its northernmost point. The area between the original course of the Rio Grande and the new course was known as the Chamizal. By terms of the original border agreement, this movement of the river placed the Chamizal and its thousands of Mexican citizens in the United States.

Neither country liked this change, and by 1963 they had negotiated a new treaty. The United States ceded much of the Chamizal to Mexico but retained the northern part. The United States' and Mexican areas are now separated by the present course of the Rio Grande, and both countries have used the areas to construct parks that are, at least theoretically, freely accessible to citizens of both the United States and Mexico.

### Nile River and Its Floods

Earlier than 3000 BC, people in Egypt knew how to live from the bounty of the Nile River. Each spring the river flooded its narrow valley (Figure 1.26), bringing water and a fresh layer of nutrient-rich silt to the agricultural fields. When the flood receded, farmers planted their crops, watered them through the summer (partly by using canals from the Nile), and harvested the crops in the fall.

This pattern of agricultural production yielded one crop per year, which was inadequate when the population of Egypt expanded rapidly in the 20th century. Partly for this reason and partly for production of electricity, Egyptians constructed a "High Dam" at Aswan. This dam controls the flooding, and Egypt can produce two crops per year by letting water through the dam at a controlled rate. The dam also impounds the river's silt in the lake behind the dam. The lack of new silt each year makes it

**Figure 1.26** White Nile, Blue Nile, and lakes in eastern Africa.

necessary for farmers to use artificial fertilizer and also causes shoreline erosion of the delta at the mouth of the Nile.

Ancient Egyptians were able to live along the Nile as far south as Aswan, but hostility of people living in what is now Sudan prevented the Egyptians from moving much farther south. A full exploration of the Nile River system did not begin until the 19th century, when European explorers and their armies began to survey much of Africa. One of their main tasks was to find the source of the Nile and understand its annual floods.

One of the earliest outposts established by the British was at Khartoum in 1821. Khartoum is very close to the junction of two branches of the Nile. The Blue Nile derives its name from the clarity of its water, whereas the White Nile is named because its high sediment load renders the water almost opaque. Khartoum remained a British possession until 1956, when it became the capital of an independent Sudan.

As early as the 17th century, explorers fairly easily traced the Blue Nile to its source in Lake Tana, Ethiopia. They also discovered that the Blue Nile contributes more than 75 percent of the water to the Nile river system and that winter rainfall in the Ethiopian highlands is the principal cause of spring flooding along the entire river system, including Egypt.

By the start of the 1860s the remaining problem was to locate the source of the White Nile. Everyone knew it was somewhere among the lakes of East Africa (see Great African Rift in Chapter 2). The problem was "which lake?"

The apparently simple method for locating the source of the river was to sail south from Khartoum. That didn't work, however, because the White Nile branches into channels that wander through swamps that were virtually impassable to vessels at that time.

Consequently, European explorers, and a few from Asia and America, went directly from the east coast of Africa to inland lakes to look for the Nile's source.

Most of them wanted the honor (and wealth) that would come from being the first person to find the source of the White Nile. John Speke, a British adventurer, claimed to have found the source in 1863 at a place where water from Lake Victoria flowed over falls into a stream headed north. The falls are now covered by a dam, and modern geographical information confirms that they are the source of the White Nile.

Many people in the 1860s, however, did not believe Speke because he had not been able to follow the Nile all the way down to Khartoum. In order to settle the issue, the Royal Geographical Society scheduled a debate between Speke and one of his antagonists for September 18, 1864. The day before the debate, however, Speke went hunting and shot himself with his own gun. No one ever knew whether his death was an accident or a deliberate suicide.

### Route Across the Great Basin

People began taking wagon trains to northern California in the early part of the 19th century, and the number increased several fold after

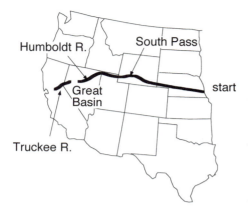

**Figure 1.27** Wagon route from Mississippi River to California, including the waterless stretch between the Humboldt and Truckee Rivers.

the discovery of gold in 1848 (see Gold in Chapter 4). Most of the travelers started from towns along the Missouri River in the spring of the year (Figure 1.27). Then they pushed westward as quickly as possible, hoping to get as far as Independence Rock, Wyoming, by July 4. Independence Rock is a large outcrop that is now a state historic site on Highway 220. Some travelers celebrated their arrival by writing their names on the rock and leaving letters to be taken home by people going back eastward.

Travelers who reached Independence Rock by July 4 could comfortably expect to arrive in California before winter snows closed passes over the Sierra Nevada. From Independence Rock the travelers moved westward toward the continental divide at South Pass, Wyoming. South Pass is at an elevation of about 2,500 meters (7,500 feet) and provides a much easier route across the continental divide than the one originally used by Lewis and Clark (see Missouri River). Land toward the western side of the pass contains numerous small rivers and gave the travelers an easy start on their way West.

The route was still reasonably easy when the wagons reached northeastern Nevada and headed down the Humboldt River. There was plenty of water in the river, and people who didn't know anything about Nevada were unaware of the dangers that lay ahead. They were entering the Great Basin.

The Great Basin is a broad area that includes almost all of Nevada plus parts of Utah and California. It contains numerous normal faults that are aligned north-south (see Chapter 2). The faults create basins that are isolated from each other as basins of "interior drainage" (see Dead Sea in Armageddon, Chapter 2). This isolation causes water that flows into the basins to be trapped, and none of the rivers that flow into the basins ever reach the ocean.

The Great Basin receives very little rain because it is in the rain shadow of the Sierra Nevada. The absence of rain prevents rivers that flow into the Great Basin from gaining any new water, and the mouths of rivers ultimately dry up when their water evaporates and leaks into the sands of the basins. Some of the basins contain dry lakes, formed by precipitation of salt in basins where rivers form a short-lived lake before drying up.

The Humboldt River flows from the east into central Nevada. In the spring, the river carries a significant amount of water from highlands in Wyoming and Idaho. That supply diminishes as the summer wears on, and the river carried very little water when the wagon trains reached it in the late summer. When the travelers came near the end of the water, they could fill barrels for themselves but couldn't possibly carry enough water for their animals. So the travelers gave the animals one last drink in the river and hoped they could make it to rivers coming east from the Sierra Nevada before the animals died.

Most of the people aimed for the Truckee River, which flows out of the Sierra Nevada into the closed basin that contains Pyramid Lake. The distance from the last water in the Humboldt River and the Truckee varied with the time of year but was commonly about 100 kilometers. Some animals didn't make it, and people were left to struggle westward with whatever meager supplies they could carry. They abandoned everything that was heavy, including furniture, excess food, and extra clothing. Even in the late 20th century, people crossing central Nevada could see bleached bones of dead animals, bottles colored by the sun, and occasionally wisps of clothing wrapped around bushes by the wind.

People who reached the Truckee River had plenty of water, but they had not yet reached their objective in California. People heading for the gold fields had to start by climbing out of the Great Basin up the steep eastern face of the Sierra Nevada. Then they had to get through narrow passes in the crest of the range before they were blocked by winter snows.

One of the passes is named after the Donner party of 87 people who came near the crest in late October 1846. When snow prevented any farther travel, the party broke up into different groups who tried to save themselves in different ways. Fewer than half of the people survived, and some of them were said to have resorted to cannibalism.

## CANALS

People have dug canals for at least 4,000 years. The oldest were apparently in the Middle East, and they are now filled with sediment and no longer used. The Chinese dug a canal linking the Yangtze and Yellow (Hwang Ho) rivers in the 4th century BC. Venice was built around canals in the Middle Ages. The Aztec capital of Tenochtitlan was cut by canals on an island in Lake Texcoco (now Mexico City). The Canal du Midi was opened in 1681 to link Toulouse, France, to the Mediterranean. The Canal du Midi provided a land route from northern France to the Mediterranean and made it unnecessary for ships to make the long voyage across the Bay of Biscay and through the Straits of Gibraltar.

Two comparatively recent canals in Germany are very important. The Kiel Canal was completed in 1895 and is heavily used for shipping between the North Sea and the Baltic Sea. The Main-Rhine-Danube canal was opened in 1992 to allow ships to pass from the Rhine River to the Black Sea.

Canal building in North America began early in the 19th century. The Erie Canal was finished in 1825. It linked the Hudson River with Lake Erie and encouraged settlers to move westward. The major transportation route between interior North America and the Atlantic Ocean was completed in 1932 when the Welland canal linked Lake Erie with Lake Ontario.

We discuss in more detail two canals that have been particularly important in history—the Suez and Panama Canals.

**The Suez Canal**

Maritime trade between Europe and Asia had to follow a route around Cape Horn until well into the 1800s. Even with faster ships of the 19th century, this route was too tedious. Some governments and companies resorted to shipping goods and people to ports on the northern and southern ends of the 150-kilometer isthmus that separates the Mediterranean and the Gulf of Suez and carting them overland between the ports (Figure 1.28).

In 1859 a private company chartered by the French government began to dig a canal across the isthmus. The route presented very few natural difficulties because it was along a geologic "zone of weakness" that extends into the Gulf of Suez from the Red Sea spreading center (see Afar in Chapter 2). Consequently the canal could be dug without locks to lift ships above sea level. Much of it was through lakes that are now merged to form the single Great Bitter Lake. The route also presented virtually no natural hazards, such as

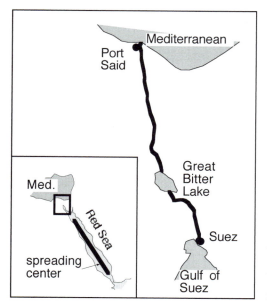

**Figure 1.28** The Suez Canal. Box in inset shows the area of the canal.

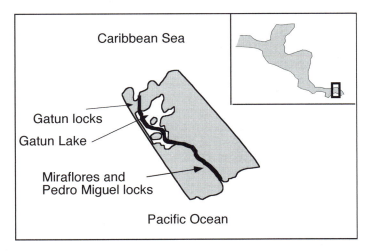

**Figure 1.29** Panama Canal. Box in inset shows the area of the canal.

earthquakes, in contrast to the seismically active zone that extends north from the Gulf of Aqaba (see Armageddon in Chapter 2).

The Suez Canal was opened in 1869 as a combined French-British venture. In 1956 Gamal Abdel Nasser overthrew the Egyptian monarchy and seized control of both Egypt and the Suez Canal. About 4 million barrels of oil pass through it everyday from producers in the Middle East to consumers in Europe.

**Panama Canal**

The Isthmus of Panama prevented easy transportation between the Atlantic and Pacific Oceans. A canal across it became increasingly necessary as the western United States became a more important part of the American economy. When Panama declared independence from Colombia, American troops immediately rushed to Panama to "protect" the new country and to start planning an American-owned canal along the easiest route across the isthmus.

Before the canal could be built, it was necessary to eradicate as many mosquitoes as possible. Malaria and yellow fever had killed more than 20,000 construction workers during an unsuccessful French effort to build a canal in the late 1800s. Doctors had known for many years that mosquitoes were responsible for spreading malaria, but their contribution to yellow fever was unclear until 1902, when it was proved by American doctors working in Cuba. One of those doctors was Walter Reed, after whom the U.S. Army Medical Center is now named.

After solving the medical problem, the next task was to find the easiest route through the mountains of Panama (Figure 1.29). To their great good fortune, surveyors located an area of relatively low elevation that was the impact crater of an enormous meteorite that had struck Panama about 20 million years ago (geologists didn't recognize that it was a meteor crater until nearly one century after the canal was finished). This area of low elevation was dammed to create an artificial lake (Gatun) that ships could sail across. Then the canal builders had to create a series of locks that could raise and lower ships between Lake Gatun and sea level on the Caribbean and Pacific sides of the canal. The whole canal was completed for ship passage in 1914, shortly after the outbreak of World War I. Ships traveling between New York and San Francisco through the canal had to cover a distance of only 10,000 kilometers, in contrast to 20,000 kilometers for the trip around Cape Horn.

Many ships built during the 1900s were deliberately designed to be narrow enough to pass through the Panama Canal, and more than 800,000 ships had used it by the end of the century. Now, however, shipping companies have found that oil and other merchandise can be moved most cheaply in very large ships, including supertankers. Most of these ships are too large to pass through the Panama Canal, and they use the old route around the southern tip of South America.

# 2

# TECTONICS

This chapter discusses processes that occur in the solid part of the earth and their effects on people. We begin by describing the nature of the solid earth, including the core, mantle, and crust, and the principle of isostasy.

We proceed to the set of principles embodied in plate tectonics. This discussion starts with the history of geologic work that led to plate tectonics, which is the realization that the earth is divided into relatively inactive plates bordered by active plate margins. We discuss the terminology and the processes that occur along the margins.

One process is difficult to fit into the concept of plate tectonics. Plumes and hot spots are very controversial among geologists. We discuss the problems, using Hawaii as an example and briefly describing other hot spots on the earth.

We follow this description of the solid earth and its processes by discussing the effects of several tectonic processes on human history and events. They include:

- Volcanism; the end of the Minoan civilization after the eruption of Santorini, and the effects of glowing ash clouds on Pompeii and Martinique.
- Earthquakes; the destruction of Mayan Quirigua; the Dead Sea fault and earthquakes at Armageddon; the San Andreas fault in California; and the 1906 San Francisco earthquake.
- Tsunamis; subduction in the Indian Ocean in 2004; and the danger of collapse at La Palma, Canary Islands.
- Living in rifts; the Great African Rift; and the extension of the mid-Atlantic ridge into Iceland.
- Living on small islands: Hawaii and Midway as examples.

## SOLID EARTH

From the center to the surface, the earth consists of a dense core, an overlying mantle, and a crust (Figure 2.1).

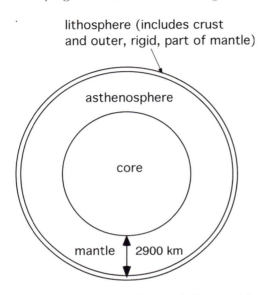

lithosphere (includes crust and outer, rigid, part of mantle)

asthenosphere

core

mantle ↕ 2900 km

**Figure 2.1** Major regions of the earth's interior.

The boundary between the core and the mantle was discovered by Swedish geophysicist Beno Gutenberg in the 1930s. He and others found that seismic waves that pass through the earth are slowed down near the earth's center, presumably because they pass through a liquid. They concluded that the top of this liquid is the core-mantle boundary and placed it at a depth of 2,900 kilometers. Because the radius of the core is approximately one-half of the radius of the earth, one-eighth of the earth's volume is in the core. The core contains more than one-half of the mass of the earth, however, because it consists mostly of a nickel-iron alloy that is denser than rocks in the mantle and crust.

The boundary between the mantle and the crust was discovered in 1909 by Croatan seismologist Andrija Mohorovicic. He noticed that the velocities of seismic waves increase suddenly below a "discontinuity" at a depth of a few tens of kilometers. He also calculated that this discontinuity was a boundary between rocks in the mantle with typical densities greater than 3 g/cc and rocks in the continental crust with densities of about 2.8 g/cc.

In his honor, we call the crust-mantle boundary the "Mohorovicic discontinuity." (Actually, geologists are intelligent enough to shorten it to MOHO.) The depth to the MOHO is variable—about 12 kilometers under oceans (combined crust and water depth), 30–40 kilometers under most continents, and 60 kilometers under mountain ranges. These depths are consistent with the principles of isostasy, which we discuss next.

Continuing geophysical studies led geologists to understand that the outer part of the mantle is different from the inner mantle. They realized that the outer part of the mantle is rigid and the inner part is mobile and flows like plastic. We now recognize a rigid "lithosphere" that combines the crust and outer part of the mantle and floats over the mobile "asthenosphere" (Figure 2.1). The asthenosphere is at the earth's surface along mid-ocean ridges (see Plates and Plate Margins) and only a few kilometers deep under continental rifts. In most continental areas, the top of the asthenosphere is about 200 kilometers deep, and it may be the surface that continental plates move upon.

### Isostasy

The floating of rigid lithosphere causes the most fundamental distinction that we can recognize on the earth's surface, which is between continents that float high on the mantle and ocean basins that float lower. The reason for this difference in elevation is that continental crust is thick and contains rocks of low density, whereas oceanic crust is thin and denser.

This balancing process illustrates the principle of "isostasy," which was first discovered when the British government financed a survey of India in the 1850s. The surveyors started in the southern part of India and worked northward to the Himalayas. This process established a "triangulated" grid across India. When the surveyors were finished, they checked their measurements by triangulating their way back to the start. They didn't get there!

Something was wrong with the survey, but the surveyors didn't know what. They knew that the Himalayas were large enough to attract a plumb bob at stations near the mountains (Figure 2.2). Therefore, the surveyors calculated the effect of the mountains on their plumb bobs by assuming that the mountains were simply set on top of the plains that make up most of India.

To find out what was wrong, the surveyors turned to two of Britain's leading mathematicians for help. G.B. Airy was professionally involved with mathematics as the Astronomer Royal of Great Britain. J.H. Pratt was more of an amateur since his primary occupation was serving as the Anglican Archdeacon of Calcutta (now Kolkata).

Both Airy and Pratt looked at the method of correction that the surveyors used. They agreed that the surveyors had overestimated the mass of the Himalayas and thus "overcorrected" their effect on the plumb bobs. But Airy and Pratt disagreed on the exact interpretation of this larger mass.

Airy proposed that the mountains were composed of the same type of rock that makes up the continental crust of the Indian plains, with the

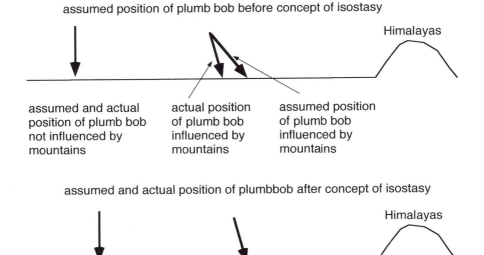

assumed position of plumb bob before concept of isostasy

Himalayas

assumed and actual
position of plumb bob
not influenced by
mountains

actual position
of plumb bob
influenced by
mountains

assumed position
of plumb bob
influenced by
mountains

assumed and actual position of plumbbob after concept of isostasy

Himalayas

**Figure 2.2** Difference between true deflection of a plumb bob by mountains that are isostatically balanced and deflection assumed before the concept of isostasy was recognized.

rock extending farther down into the mantle beneath the mountains. Pratt assumed that the mountains consisted of lighter rock than the plains, and the mantle beneath both of them was at the same depth.

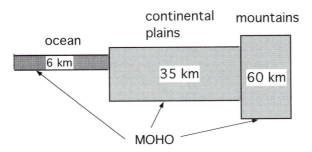

**Figure 2.3** Depths of MOHO below isostatically balanced crust.

Both the Airy and Pratt hypotheses signify that all elevations on the earth are controlled by the thickness and density of the underlying crust and are, thus, of "equal standing" (Figure 2.3). We call this concept "isostasy" and say that the earth's surface is "isostatically balanced." Geologists explain this balance by a combination of the Airy and Pratt models. Other geophysical methods show that Airy was correct for mountains because they clearly have light roots that extend into the mantle. On the other hand, ocean basins are lower than continents because their crust is both thinner and denser than in the continents.

## PLATE TECTONICS

The earth's crust is constantly in motion as a result of heat generated in the earth's interior. It causes movements of lithospheric plates and also "hot spots," which may be areas of rising mantle known as "plumes." We discuss the heat before we describe tectonics and plumes.

### Earth's Heat

Geologists of the 19th century made the first tentative efforts to understand deformation of the earth's surface, a process that we now call "tectonics." They noticed that mountain belts were places where the earth had been compressed and thought the shortening might have been caused by shrinkage of the earth. They compared the process to the drying out of a prune, with mountain belts as the places where the skin had been wrinkled.

Geologists thought the earth was shrinking because it was cooling off. They knew that the interior of the earth was hot because they saw volcanoes erupting. They also could calculate the heat produced when the earth accreted from the solar system. This heat was equal to the potential energy lost when particles from the outer part of the solar system crashed into the growing earth, and geologists knew that it wasn't enough to keep the earth warm for more than a few million years.

The idea that the earth was cooling and shrinking came to an end in 1896. In that year, French scientist Antoine Henri Becquerel accidentally left photographic film in a closed drawer with some fragments of uranium ore. When he found that the film had been exposed to radiation in the drawer, he knew that the exposure was not caused by sunlight but by some new form of radiation. He called it "radioactivity."

Geologists knew that the earth contained uranium and thorium (later found to be radioactive), and now they knew that the energy released by radioactive decay was enough to keep the earth hot enough to prevent shrinking. The recognition of a continued source of heat in the earth led to two important conclusions. First, it enabled geologists to realize that the earth is much older than they thought it was (see Chapter 3). Second, it let geologists understand that the earth's heat provides enough energy to drive tectonic processes.

### Structures and Processes

Geologists of the 19th century recognized most of the fundamental structures that are used to describe movement of rocks. They found

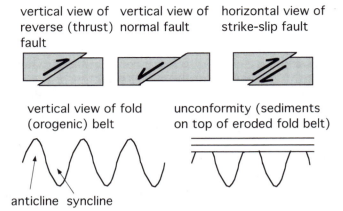

**Figure 2.4** Types of faults and folds; diagram of unconformity.

that the earth's surface is cut by three types of faults. Using modern terminology, these faults are identified as "normal" where the earth's surface is stretching, "reverse" or "thrust" where the earth's surface is being compressed, and "strike-slip" where parts of the surface are moving horizontally past each other (Figure 2.4). These three types of movement would later be used to classify plate boundaries.

These early geologists also saw that some rocks had been folded (Figure 2.4). Folds are classified as anticlines, where rocks have been pushed upward, and synclines where they have been pushed downward. Anticlines and synclines commonly form a series of folds parallel to each other. These "fold belts" and accompanying thrust faults are zones of compression that are mostly the remains of old mountain belts. We refer to them as "orogenic belts" and the process that formed them as "orogeny." After the end of an orogeny, folded rocks are commonly eroded to form an "unconformity" that is the base of deposition of flat-lying sediments.

### Continental Rifting

The idea of continental rifting was first published by British geologist J.W. Gregory in 1896. He studied the rift in Kenya, where land rises on both sides to the margins of the rift and then suddenly drops away into the Great African Rift (Figure 2.5) The floor of the valley is a few hundred meters below the rim and the valley is about 50 kilometers wide. Several volcanoes, including Longonot (Figure 2.6) occupy the rift.

We now know that the Great African Rift system began to develop at about 30 Ma. The rift system radiates from Afar, which is a hot spot, possibly the center of a plume (see Hot Spots and Mantle Plumes). The rift extends southward from Afar and then divides into a branch that passes east of Lake Victoria and one that passes to the west. The eastern branch is regarded as "wet" because it is filled with volcanoes like Longonot that erupt large volumes of smoothly flowing basalt. Conversely, the western branch is regarded as "dry" because it contains only a few volcanoes that erupt infrequently but are explosive and dangerous.

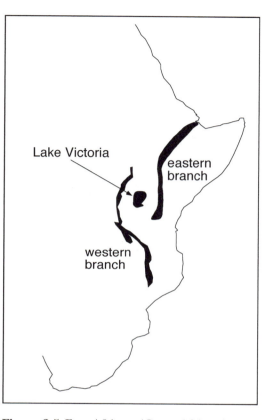

**Figure 2.5** East African (Great African) rift.

### Deep Seismic Zones

In 1954, Caltech geophysicist Hugo Benioff published seismic data that showed places where the earth's lithosphere (crust and upper mantle) plunged down to a few hundred kilometers into the mantle along steeply inclined planes of earthquake activity. Benioff was repeating work published in the 1930s by Japanese geophysicist K. Wadati. Wadati's work was forgotten during World War II, however, and the inclined planes were called "Benioff zones." We now call them "subduction zones," using a term first published in 1974 by D.H. White and three other American geologists (Figure 2.7).

### Mid-Ocean Ridges and Topography on the Ocean Floor

The first evidence that the ocean floors were not simply smooth surfaces came in 1866 when a Trans-Atlantic telegraph cable was laid

**Figure 2.6** The Longonot Volcano in the African rift in Kenya.

between Ireland and Newfoundland. Much to the surprise of the people laying the cable, it crossed a broad ridge approximately in the middle of the ocean between deep plains on each side.

Further information about oceans was collected when the British sent out the research vessel *Challenger* to learn as much as possible about the world's oceans. The *Challenger* cruised around the world from 1872 to 1876 and collected so much information that it virtually established the field of oceanography.

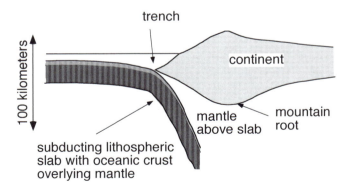

**Figure 2.7** Major features of continental-margin subduction zone.

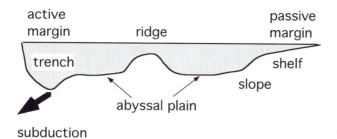

active margin    ridge    passive margin

trench    shelf

slope

abyssal plain

subduction

**Figure 2.8** Active and passive continental margins.

The *Challenger* expedition discovered that all oceans contain ridges with typical depths of 1 to 2 kilometers between "abyssal plains" with average depths of 5 kilometers. The expedition also discovered two very different kinds of continental margins (Figure 2.8). On some margins the continents extend outward up to several hundred kilometers as shallow "continental shelves" until they reach depths of about 200 meters, where slightly steeper "continental slopes" connect the shelves with the abyssal plains. We now call these margins "passive" because no earthquake or volcanic activity occurs along them.

The *Challenger* expedition also discovered that some continental margins and all margins of island arcs plunge abruptly to great oceanic depths. Many of these margins also have deep "trenches" where some depths are greater than 10 kilometers. We now call these margins "active" because they are the sites of earthquakes and volcanism.

Much of the confusion about the nature of the ocean floor was cleared up in 1960 by Princeton geologist Harry Hess. In what he referred to as "an essay in geopoetry," Hess suggested that mid-ocean ridges were places where new oceanic crust formed as the crust on opposite sides pulled apart (we now call them "accretionary margins"). This process was called "sea floor spreading" by American geologist Robert Dietz in 1961.

Confirmation of seafloor spreading was first published in 1963 in a paper by F.J. Vine and D.H. Matthews of Cambridge University. They knew that the magnetic field of the earth occasionally reverses itself ("flip-flops"), putting the north magnetic pole near the south rotational pole and the south magnetic pole near the north. Time intervals have "normal" polarity when the north magnetic pole is in the north and "reverse" polarity when the north magnetic pole is in the south.

Vine and Matthews also knew that rocks that contain iron inherit a magnetic orientation from the earth's magnetic field when they form.

If old rocks formed during a normal interval, then they enhance the present earth's magnetic field near them. Rocks formed during reverse intervals interfere with the present magnetic field and make it weaker near them. Vine and Matthews reasoned that the iron-bearing basalts of the ocean floor would affect the present earth's magnetic field measured on ships cruising over mid-ocean ridges. The result was the discovery of "magnetic stripes" (Figure 2.9).

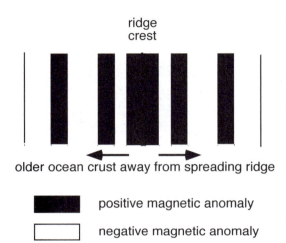

ridge crest

older ocean crust away from spreading ridge

■ positive magnetic anomaly

□ negative magnetic anomaly

**Figure 2.9** Magnetic stripes around mid-ocean spreading center.

These stripes form because basalts erupted on ridge crests inherit the orientation of the earth's magnetic field at the time of eruption and then carry these old orientations with them as they spread farther away from the active ridge crest. Where this orientation is in the same direction as the present magnetic field, the strength of the field is stronger, and the stripe is said to have a "positive anomaly" (Figure 2.9) Conversely, rocks whose magnetic orientations are in the opposite direction to the present field have weaker fields and are said to have "negative anomalies."

Geochronologic investigation later confirmed that the oceanic basalts are older the farther they are from the ridge crests. Because the basalts form along mid-ocean ridges, they are referred to as "mid-ocean-ridge basalt (MORB)." This evidence of modern volcanism at ridge crests and older rocks on the sides finally convinced almost everyone of the reality of continental drift and seafloor spreading.

The idea that mid-ocean ridges cause oceans to expand by creating new lithosphere led geologists to rethink many of their earlier concepts of earth history. Spreading provided a mechanism for continental drift, which was immediately accepted by all but a recalcitrant few geologists (see Pangea in Chapter 3). Creation of new lithosphere also required that lithosphere be destroyed somewhere else unless the earth is expanding, which it clearly isn't. This realization also explained that passive continental margins occur on the sides of oceans that are expanding, whereas active margins develop around oceans that are contracting.

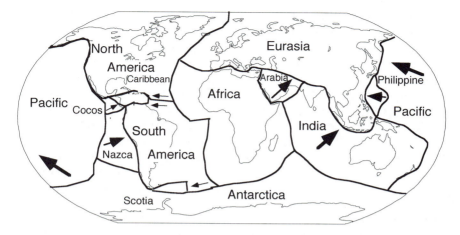

**Figure 2.10** The earth's current plates. Arrows show directions of movement.

The idea of spreading also led geologists to recognize the profound differences between the Pacific Ocean and the other oceans on the earth. The Pacific Ocean is shrinking as the Atlantic, Indian, Arctic, and Antarctic Oceans expand. Because the shrinkage of the Pacific Ocean is caused by subduction, numerous earthquakes and volcanoes occur along the rim of the ocean. It is popularly known as the "ring of fire."

### Plates and Plate Margins

All of the information about tectonic processes was gathered together by geologists in the late 1960s and early 1970s and called "plate tectonics." Although not a process in itself, the concept of plate tectonics gave geologists an opportunity to think more clearly about the earth than they ever had before. Within 10 years after the discovery of seafloor spreading, geologists had outlined all of the world's plates. The earth's surface is now covered by seven large plates and several smaller ones (Figure 2.10). Many of the plates, such as South America, contain both continental and oceanic crust, which means that some continental margins are along subduction zones and some continental margins are "within plate."

Plate tectonics classifies plate margins into three types:

• Margins where plates move ("rift") away from each other and commonly form normal faults. In the oceans these margins are called "mid-ocean ridges," "spreading centers," or "accretionary margins" (Figure 2.9). Rifts in continents are simply referred to as "rift valleys" (see Continental Rifting), and the amount of separation is much smaller than in ocean basins.

- Margins where plates move toward (collide with) each other and commonly form thrust faults. They are "destructive margins" where present oceanic lithosphere forms a slab that is thrust down (subducted) into the mantle (Figure 2.7). Subduction can occur both under continental margins and within ocean basins. Subduction in both locations causes molten material ("magma") to rise from the earth's interior and solidify when it cools either within the crust as "intrusive" rocks or after it is erupted from volcanoes on the surface (see Volcanic Eruptions). Subduction creates different features within ocean basins than on continental margins. Subduction within ocean basins causes volcanism that forms "island arcs." Subduction beneath continental margins causes compression and is responsible for creating most of the world's mountain chains both now and in the past. This orogeny is accompanied by lateral shortening and vertical thickening of the crust.

- Margins where plates are moving horizontally past each other. Long zones of lateral movement are commonly known as "transforms," and short zones of movement are simply called "strike-slip faults" (see Armageddon and San Andreas).

### Hot Spots and Mantle Plumes

One important tectonic process may be unrelated to plates and plate margins. Some places on the earth appear to have been volcanically active for tens of millions of years. They are referred to as "hot spots and create areas tens of kilometers in diameter where volcanism is intense. Some of the hot spots are places where enormous volumes of basalt have been erupted to form basalt "plateaus." Some hot spots seem to have remained in place as plates passed over them, forming hot spot "tracks" more then 1,000 kilometers long. Some geologists regard the hot spots as places where "plumes" of asthenosphere rise close to the surface.

Hot spots and plumes were initially proposed by Canadian geologist J. Tuzo Wilson in 1963. They were later more fully described by Princeton geologist W. Jason Morgan in 1972 to explain the pattern of the Hawaiian Islands and features to the northwest of them. We discuss Hawaii first and then other hot spots and the basalt plateaus associated with them.

### Hawaii

A chain of volcanoes and volcanic islands stretches from Loihi, in the southeast, to Meiji, in the northwest (Figure 2.11). Loihi is a new volcano that is growing but has not yet emerged above the sea. Northwestward of Loihi is the "Big Island" of Hawaii. It contains Kilauea, which is

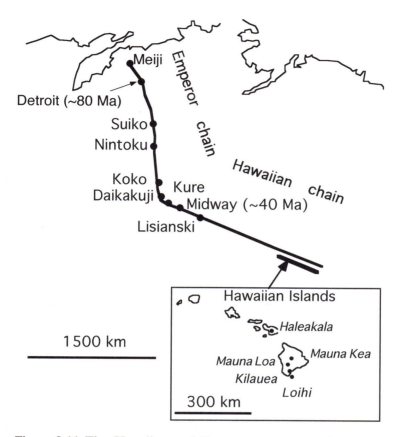

**Figure 2.11** The Hawaiian and Emperor seamount chains are the thick line. Hawaiian Islands (in box) are the southeastern part of the chain. Some of the seamounts are shown as black dots (with ages of Midway and Detroit in Ma). Small black dots show major volcanoes in Hawaiian Islands.

violently active, and the intermittently active Mauna Loa and dormant Mauna Kea. The next island to the northwest in the chain is Maui, where Haleakala volcano was most recently active a few hundred years ago. Farther northwest are islands formed by now-extinct volcanoes, including Oahu, the site of Honolulu, and Kauai, the last island in the State of Hawaii.

Continuing northwestward beyond Kauai, the islands change remarkably. At first there are only islands formed by the small tips of volcanoes that barely rise above sea level. Farther northwestward are islands where volcanoes have subsided below sea level. The only visible parts are atolls consisting of circular reefs that remained at sea level and grew upward and outward from their volcanic base as the volcanoes subsided. The reefs surround lagoons that contain small islands where storms tossed reef debris into the lagoons, and waves were able to build some of it above sea level. The largest of these atolls is Midway, and about 2,000 kilometers northwest of Hawaii is the tiny atoll of Kure (Ocean Island).

The volcanic islands and atolls from Hawaii to Kure are created by volcanic prominences on a submarine ridge. Some of the prominences are completely submerged and are referred to as "seamounts." They give the name "Hawaiian seamount chain" to the underlying ridge.

A similar ridge continues northwest from Kure at an angle of about 30° to the Hawaiian seamount chain (Figure 2.11). Volcanic prominences on this ridge, however, are completely submerged and the ridge is referred to as the "Emperor seamount chain" because most of the extinct volcanoes are named after Japanese emperors. The northernmost seamount is Meiji, just south of the intersection of the Aleutian trench with Kamchatka. There may once have been older seamounts along the chain, but geologists have no knowledge of them because they would have been destroyed when the Pacific plate was subducted either beneath Asia or the Aleutians.

The existence of seamounts and their presence throughout much of the Pacific Ocean was first learned from oceanographic research begun during World War II. One of the people who discovered seamounts was Harry Hess, who suggested "guyot," the name of the Geology building at Princeton, as an alternative term for seamounts. Even before most of the seamounts, islands, and the volcanic bases of atolls were dated, it was clear that the age of volcanism increased as the distance from Hawaii increased. This early conclusion has now been strengthened by radiometric dates shown in Figure 2.11. This pattern of lower elevation with increasing age occurs because volcanic peaks cool off as they age, and older ones (farther from Hawaii) have subsided more than peaks near Hawaii. Geologists who regard Hawaii as the site of a plume explain the Hawaii-Emperor seamount chains as a plume track.

*Other Active Hot Spots*

Geologists have proposed that the world contains more than 50 hot spots, and Figure 2.12 shows the most important ones. We discuss Yellowstone in Chapter 1 and restrict ourselves here to Afar, Iceland, and Reunion Island).

*Afar.* The Afar hot spot is currently under the country of Djibouti at the southern entrance to the Red Sea (Figure 2.12). Volcanism in

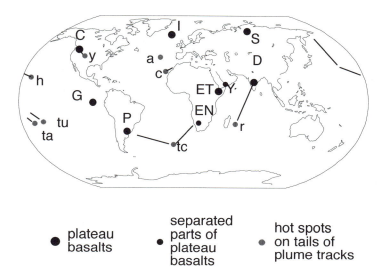

Figure 2.12 Hot spots. Abbreviations of plateau basalts are:
C, Columbia River; D, Deccan; ET, Ethiopia; G, Galapagos;
I, Iceland; P, Parana, S, Siberia. Abbreviations of separated
parts of plateau basalts are: EN, Entendeka; y, Yemen. Abbreviations of hot spots on the tails of plume tracks are: a, Azores;
c, Canary Islands; h, Hawaii; ta, Tahiti; tc, Tristan da Cunha;
tu, Tuamoto. (The Tuamotos are a chain of coral reefs that
are presumed to represent a plume track although there is no
active volcanism in them.)

Afar began at about 30 Ma and was preceded slightly by eruption of the
enormous Ethiopian basalt plateau. Opening of the Red Sea split the
plateau into two parts, with almost all of it remaining in Ethiopia and
only a small part on the opposite shore in Yemen.

Three rifts extend from Afar, and it is regarded as a "triple junction"
(Figure 2.13). One rift is the Great African Rift discussed earlier (Figure 2.5). A second rift is the northwestward extension of the Indian
Ocean ridge, which caused much of the opening of the Indian Ocean.
The third rift extends northwestward as a spreading center responsible for opening of the Red Sea. Because two of the rifts that radiate from Afar caused ocean basins to open, and the African rift system did not, geologists refer to the Great African Rift as a "failed
arm."

*Iceland.* Iceland is a hot spot intersected by two spreading ridges. The
oldest rocks in Iceland are completely buried by modern volcanic rocks,

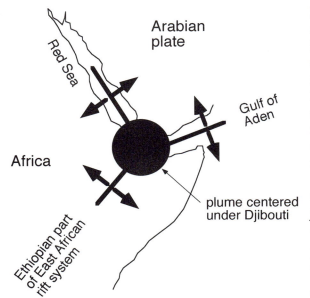

**Figure 2.13** Triple junction at Afar.

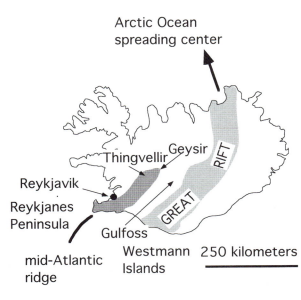

**Figure 2.14** Iceland.

but geologists assume that volcanism began at about 40 Ma, when Greenland separated from Europe. Although most of the volcanic rocks in Iceland are basalts, it differs from other hot spots because more than 10 percent of the volcanic rocks are high-silica rhyolites.

The mid-Atlantic ridge rises to the surface along the Reykjanes peninsula in southwestern Iceland (Figure 2.14). It extends northeastward past Reykjavik through a hilly area to the well-displayed rift at Thingvellir. A "transfer zone" joins the northern tip of Thingvellir to the Great Iceland Rift.

The Great Iceland Rift is the southernmost extension of the Arctic Ocean spreading center, which caused the opening of part of the Arctic Ocean. The Great Rift is volcanically active both in the center and along the rift flanks. The most spectacular modern volcanism is at Hekla, which has been regarded as a major cause of a long cold period in Europe about 10,000 to 12,000 years ago.

Volcanism along the Great Rift also continues south of Iceland in the Westmann Islands. In 1973 an eruption on the island of Heimaey covered part of the local town and almost closed the entrance to the harbor; it was stopped when the Icelandic government sent boats to use fire hoses to chill the flow by pouring water on it. Also in 1973 a new island, Surtsey, emerged above the ocean to join the Westmann group.

*Reunion Island.* Reunion Island, in the western Indian Ocean, is nicknamed "island with big spectacle." The spectacle is generated by the volcano Piton de la Fournaise. It rises to a height greater than 2,500 kilometers and has been almost continually active in historic time. The island also contains an extinct volcano, Piton des Neiges, which is more than 3,000 meters high.

Reunion is the southernmost island on the Mascarene ridge. We can trace the Mascarene ridge and its apparent extension to the north back to an enormous area of basalt in India. This Deccan "basalt plateau" was erupted at 65 Ma and may have been the head of a plume that has now dwindled to a small remnant under Reunion Island.

*Basalt plateaus.* Initial eruptions at hot spots generate large volumes of basalt and are referred to as "basalt plateaus" if they occur on land. We can recognize five of them (Figure 2.12). Two of the eruptive centers seem not to have moved after they were formed. They include the Ethiopian plateau and the Siberian basalts (see Chapter 3). Three major basalt plateaus, however, were the initial volcanic outpourings that led to plume tracks.

Comparison of the volume of basalt in plateaus with the much smaller volume of material erupted along plume tracks leads to the conclusion that basalt plateaus are erupted from the heads of plumes and the tracks from the plume tails. This pattern is shown by the track from the Deccan basalts of India to Reunion Island, from the Parana basalts of South America and the Entendeka basalts of Africa, and from the Columbia River basalts to Yellowstone.

## EFFECT OF TECTONICS ON HUMAN HISTORY

We divide our discussion of the effect of tectonics into five topics: volcanic eruptions that were too small to affect climate (see Chapter 1 for climatic effects); earthquakes; tsunamis; living in rift valleys; and living on small volcanic islands.

### Volcanic Eruptions

Molten rock known as "magma" forms when rock in the crust or shallow mantle melts. Most magmas form as "partial melts" of the lowest-melting components of a source rock. Melting occurs when the temperature of the rock is above the melting temperature, either because some process raises the actual temperature or because water in the rock lowers the melting temperature.

Magmas rise through the crust because the liquid has a lower density than the rock around it. Magmas begin to solidify, mostly by crystallization of minerals, because they cool off as they rise. Some magmas crystallize within the crust and are known as "intrusive" rocks. Some magmas, however, are erupted on the earth's surface as "extrusive rocks," either as "lava" that flows over the surface before it solidifies, or as ash blown explosively out of a volcano.

Most volcanic eruptions are not dangerous. Basalts erupted along oceanic spreading centers and in hot spots generally flow smoothly from the fissures. Volcanic rocks erupted above subduction zones, however, are from magmas that formed by melting complex mixtures of source rocks. Consequently, the subduction-zone volcanoes erupt a range of products from smoothly flowing basalt to explosive ash.

Explosive eruptions can be described partly by their "volcanic eruptivity index" (VEI) on a logarithmic scale from 1 to 8. The VEI is a measure of the amount of material blown explosively out of a volcano during one sustained phase of eruption. It can be measured directly for active volcanoes but must be estimated from the amount of preserved debris for past eruptions. Examples of the VEI include: probably 8 for Toba and the Krakatoa eruption of 536 AD (see Chapter 1); 7 for Tambora in 1815 (see Chapter 1); 6 for Santorini; 5 for Vesuvius, and 4 for Mount Pelée.

We describe three historically significant eruptions here—Santorini and the glowing cloud eruptions of Vesuvius, and Mount Pelée.

### Santorini and the End of Minoan Civilization

The landing for ships at the tourist island of Thera is inside the bowl ("caldera") of an enormous volcano (Figure 2.15). Radiometric dating shows that Thera is one of the remains of a volcano (named Santorini) that exploded in 1640 BC. The explosion was so violent that it destroyed almost all of the formerly large island of Santorini, leaving small islands as fragments of the original island (Figure 2.16). The center of Santorini is now filled by water of the Mediterranean Sea, and the possibility of further eruption is shown by continued growth of the small island of Nea Kameni inside the caldera.

From the top of the cliff above the landing, Thera slopes gently toward the eastern shore. Near the south coast is an archaeological site known as Akrotiri (Figure 2.16). It was a remarkable town. The streets were well laid out, and even had a plumbing system. Akrotiri seems to have been

**Figure 2.15** Western cliff of Thera, inside the Santorini caldera.

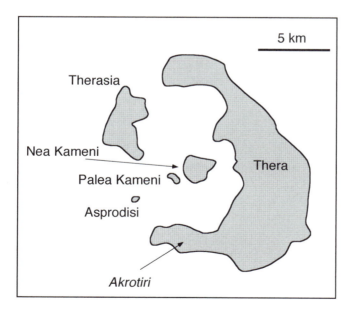

**Figure 2.16** Map of Santorini.

occupied until 1640 BC as one of the principal cities of a civilization known as Minoan.

We know a little about Minoan culture, partly from Akrotiri but mostly from the palace of Knossos and other large buildings on Crete. Early Minoans used a hieroglyphic form of writing that no one today can read, and early Greek records indicate that the early Minoan language was not from the same linguistic group as Greek. The Minoans liked decorations, both on wall frescoes and on pottery. Their pottery is recognizable by its common octopus motif. Wall paintings show that their major sports were boxing and bull jumping, with men and women participating in both of them. Above all, the Minoans were a seafaring people who engaged in trade all around the eastern Mediterranean, including the European and African coasts.

The Minoans apparently arrived in the Mediterranean at some time between 3000 and 2000 BC. Certainly by 2000 BC their trade routes can be mapped by the types of goods exchanged between Europe (now Greece), Africa, Asia Minor, and the Mediterranean islands.

The Minoan culture seems to have flourished for only a few hundred years, and by about 1500 BC, the Minoans were being supplanted by new arrivals in Greece. These Mycenaeans spoke a language like modern Greek, wrote in a script like modern Greek, and were presumably the ancestors of modern Greeks. By 1000 BC, the Mycenaeans were clearly the dominant power on the Greek mainland and throughout the Aegean Sea.

There are several possible reasons for replacement of Minoan civilization by the Mycenaeans. Possibly the Mycenaeans were better warriors, both on land and sea. Possibly the Minoans had been such shrewd traders that they had worn out their welcome in the eastern Mediterranean and were forced out by a coalition of people. Possibly Santorini exploded.

An explosion at 1640 BC is remarkably close to the time of the decline in Minoan civilization. We also know that the Minoans knew about the impending explosion and took steps to protect their people. Not a single skeleton has been recovered from Akrotiri.

### Glowing Clouds

Glowing clouds develop from magmas that are sticky because of their high concentration of silica and their abundance of water. Lavas of this type cannot simply flow down the sides of volcanoes, and they break up into a mixture of liquid droplets, solid ash, and superheated steam. They move away from volcanoes at typical speeds of 100 kilometers per hour and temperatures more than 400°C. Glowing clouds typically bury and incinerate everything in their paths. We discuss two of the most famous ones—Mount Vesuvius and Mount Pelée.

*Mount Vesuvius.* No place illustrates the proverb "Pride goeth before a fall" better than the Roman town of Pompeii.

A city of some 20,000 residents south of Rome, Pompeii was a model of luxury in the 1st century AD for Roman citizens who wanted to get away from the crowded capital. Wealthy Romans had vacation villas nearby, and the neighboring area was rich in agriculture, including grapes for wine. Contemporaneous records and archaeological information show that Pompeii had houses, stores, and streets made by close-fitting stones. At its peak, Pompeii had a large traditional Roman forum, 25 street fountains, 4 public baths, a swimming pool, and numerous frescoes in both public places and private houses. Pompeii was also a port at the mouth of a river, although recent sedimentation has now moved the shoreline away from the ruins of Pompeii.

Pompeii was idyllic for rich Roman citizens, but that happiness was built on the labor of slaves. As Roman legions expanded their empire, they brought slaves back from the newly conquered lands. These slaves, both men and women, were put to work to make life easier for Romans. A few of the athletic slaves provided entertainment as gladiators.

The privileged citizens of Pompeii were apparently unconcerned that their prosperity was based on slavery. And they didn't worry too much about the frequent minor earthquakes in the area and simply began to rebuild after a quake in 62 AD caused significant damage. The residents also presumably didn't worry that Pompeii was only about 5 kilometers away from Mount Vesuvius (Figure 2.15), which had never erupted in historic time.

That inactivity came to an abrupt end about noon on August 24, 79 AD (date adjusted to modern calendars). On that day Vesuvius erupted about 1 cubic mile of white-hot ash during a period of nearly 24 hours. During the first 6 hours, the volcanic fragments rose as high as 10 kilometers in the air before they fell to the ground as hot ash and some larger particles. Many of the inhabitants either fled on land or were rescued along the shore by ships of the Roman navy.

A few thousand people chose not to flee but to seek shelter in their houses. They were trapped a few hours later when the sides of Vesuvius collapsed and glowing clouds cascaded down the flanks of Vesuvius. Recent archaeological work has shown that several thousand people were killed by these clouds.

The town of Herculaneum, which was smaller and closer to Vesuvius than Pompeii, was completely buried by ash, and Pompeii was so covered that it could not be revived as a city. The only people who returned before archaeological work began in the 1700s were looters. After the ash cooled, some people tunneled through it to recover jewelry and other precious objects.

Vesuvius has erupted some 50 times since 79 AD, all of them smaller than the one that destroyed Pompeii. The eruptions and earthquakes are now monitored closely by Italian geologists in an effort to predict future disasters. The city of Naples, with a population of about 1 million, is only a few kilometers farther from Vesuvius than Pompeii.

*Mount Pelée.* The island of Martinique is a French Department in the Lesser Antilles island arc on the eastern edge of the Caribbean. The arc is underlain by subducting lithosphere of the Atlantic Ocean, and all of the islands were formed by volcanism and growth of reefs. Martinique was built around six volcanoes that range in age from 50 Ma to the present. The only volcano active in historic time is Mount Pelée, which is about 10 kilometers north of the town of St. Pierre. Pelée let off massive explosions of steam in 1792 and 1851 but then remained quiet until April 24, 1902. During the next week, Pelée emitted water vapor and ash that covered much of Martinique. Eruptions accompanied by loud noises became more violent on May 4, and residents of St. Pierre saw a fiery glow above the volcano by May 6. The situation was so frightening that on May 7 the mayor announced that the government had everything under control and the citizens had nothing to worry about.

Wrong! Pelée blew up at 8:00 AM on May 8. Observers in surrounding hills reported that a glowing cloud nearly 500 meters high crashed into St. Pierre. It covered or burned the entire town and killed about 25,000 people, including the mayor. The only known survivor was a convict who was in an underground cell in the town's prison. After his release, he was called the "survivor of St. Pierre" and toured the United States with P.T. Barnum's circus.

### Earthquakes

Earthquakes happen when faults slip suddenly. They occur almost entirely in subduction zones and along strike-slip faults, which show the greatest resistance to movement, and they are rare along normal faults because the opposite blocks move away from each other.

Friction along fault planes causes rocks on opposite sides to bend before they break along the fault. This slippage releases the energy that was built up by the bending before the blocks on opposite sides could move (Figure 2.17). The energy released as the bends straighten out is dispersed from the fault as various types of "seismic waves." They radiate away from the center of movement, which is called the "focus." The focus is at different depths for different earthquakes, and the point on the surface directly above it is the "epicenter." The ground displacement caused by the seismic waves is greatest at the epicenter and diminishes as the waves spread out.

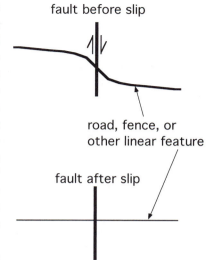

Ground displacement and the amount of energy released are related but are measured on different scales. "Intensity" is a subjective measure of

**Figure 2.17** Accumulation and release of strain along fault.

ground displacement and is commonly described by the "Modified Mercalli Scale," which was first developed in 1902 by Italian geophysicist Giuseppi Mercalli. The scale is from I to XII, with an intensity of I meaning an earthquake that was barely noticed and intensity of XII meaning total destruction.

A more quantitative scale is "magnitude," which is the amount of energy released by the fault slippage and is calculated from the sizes of seismic waves measured on seismograms. It was designed by Caltech geophysicist C.F. Richter and is reported on a scale with a lower limit of 0 and no upper limit. Differences between each number on the scale are factors of 30 in energy release. For example, a magnitude 5 earthquake releases 30 times as much energy as one of magnitude 4. Magnitude does not depend on distance from the epicenter because it is a measure of energy released at the focus.

The highest magnitude ever measured is 9.5 for an earthquake in Chile on May 22, 1960. An earthquake with a magnitude of 9.0 at Anchorage, Alaska, in 1964 was the largest ever measured in North America. The famous San Francisco earthquake of 1906 had a magnitude of "only" 8.0. For comparison, the atomic bombs dropped on Hiroshima and Nagasaki in 1945 released energy equal to a magnitude of about 5.0, but they had virtually no seismic effect because they exploded in the air.

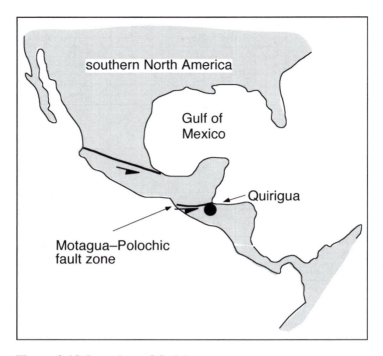

**Figure 2.18** Location of Quirigua.

We discuss the probable effects of earthquakes at Mayan Quirigua, possible ones at the Israeli site of Armageddon, and the San Francisco earthquake along the San Andreas fault of California.

### Earthquake at Mayan Quirigua

The elaborate Mayan empire began to replace the Olmec civilization in eastern and southern Mexico in the first few centuries AD. The Mayans and other people to the West flourished until Spaniards conquered Mexico in the 1500s and nearly destroyed all of the existing civilizations.

The Mayan city of Quirigua was established about 550 AD, and its location along the Motagua River made it a trading center where jade and obsidian from the Guatemalan highlands were exchanged for products from the Caribbean coast (Figure 2.18). The location seemed ideal until about 800 AD. Something obviously happened then because there is no record of activity in Quirigua after that time, and the city is far more destroyed than any of the other large Mayan sites.

Quirigua is along the Motagua-Polochic fault zone, which has been seismically active in historic time (Figure 2.18). The evidence that an earthquake around 800 AD destroyed Quirigua is not hard to find—collapsed buildings, broken and cracked stones and mortar, places where tall stone slabs (stelae) had fallen over. The destruction was even more severe than visitors think it was because some of the stelae had been picked up and restored to vertical in 1934 when people tried to make Quirigua look as much as possible as it had looked before it was destroyed.

### Armageddon

One of the world's longest faults runs from the Gulf of Aqaba through the Dead Sea, the Sea of Galilee, and ultimately ends in the Taurus Mountains at the southern border of Turkey (Figure 2.19). This Dead Sea fault is called a "transform" (see Plates and Plate Margins) because it is a line of movement that separates two plates with very different geologic histories.

Movement on the Dead Sea transform began at about 18 Ma. At this time the eastern side (Arabia) began to move northward, and total movement since then has been more than 80 kilometers. Movement of the block east of the transform closed the eastern end of the Mediterranean, which had formerly connected to the Indian Ocean, and formed the Taurus Mountains by driving the Arabian Peninsula into Turkey.

The basin occupied by the Dead Sea is in an area where the stresses that created the transform caused down warping by some combination of faulting and folding. Because the Dead Sea basin is below sea level, water can only flow into it and not out of it. The only inlet is the Jordan River, whose water is being used so much for agriculture that the Dead Sea has shrunk to just slightly more than half of the size it had 25 years ago. People who go to the tourist areas of the Dead Sea can put on bathing suits and float on the water if they wish, but they need to wash off thoroughly with fresh water when they get out to prevent their skin from being irritated by dried salt.

The Dead Sea transform is, like all other transforms, an area of earthquake activity. At least one earthquake destroyed the ancient city of Petra, in Jordan. Petra is a tourist attraction because so many of its buildings were carved out of solid rock. It was built by a group of people referred to as Nabateans, and they used Petra as a center to control trade routes between the Mediterranean and the desert lands of the Arabian Peninsula during a period of several hundred years. Buildings in Petra show patterns of destruction that could have been caused by an earthquake, and it is likely that the city was abandoned after an earthquake in 363 AD made it uninhabitable.

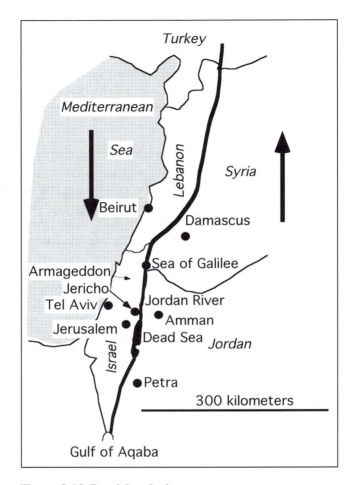

**Figure 2.19** Dead Sea fault system.

The area north of the Dead Sea is a location of frequent earthquakes. This frequency is created by the bend in the transform (Figure 2.19). Because of the bend, the eastern and western plates cannot push smoothly past each other, and the increased friction causes more earthquakes. Geologists call this feature a "restraining bend."

Numerous earthquakes have been recorded near Jericho, and the sites of other earthquakes are north and west of Jericho. One location is at a small hill about 50 meters high known as "Har Megiddo" (hill of Megiddo). It is clearly a place of repeated earthquake activity, and archaeologists know that cities near it were destroyed at least five times and then rebuilt over the ruins of the former cities.

Not all of the destruction along the Dead Sea transform may have been caused by earthquakes. Just north of the Dead Sea, Joshua "crossed the Jordan" and led the Hebrews into the "promised land" between the Jordan and the Mediterranean Sea. Archaeologists are not sure whether the destruction of Jericho was caused by an earthquake or by Joshua's army because it is very difficult to distinguish the results of a large earthquake from the sacking and looting carried out by ancient armies. And the Book of Joshua makes it clear that the city was looted (chapter 6, verses 20 and 21): *20, When the trumpets sounded, the people shouted, and at the sound of the trumpet, when the people gave a loud shout, the wall collapsed; so every man charged straight in, and they took the city; 21, They devoted the city to the Lord and destroyed with the sword every living thing in it—men and women, young and old, cattle, sheep and donkeys.*

Har Megiddo poses a more famous puzzle about earthquake destruction. It sits astride the main route for ancient people traveling between Assyria, to the north, and Egypt to the south. This strategic position made it a site of numerous battles involving Assyrians, Egyptians, and different tribes of Hebrews. With all of this carnage in its history, it is hardly surprising that many people think Har Megiddo is really Armageddon, whose adjacent plain is thought to be the site of a great battle between the forces of good and of evil.

Biblical prophecy of an earthquake at Armageddon occurs in the Book of Revelation (chapter 16, verses 16 and 18): *16, And he gathered them together into a place called in the Hebrew tongue Armageddon; 18, And there were voices, and thunders, and lightnings; and there was a great earthquake, such as was not since men were upon the earth, so mighty an earthquake, and so great.*

The possibility that Armageddon is also the battle between God and the forces of evil is in the book of Joel, which does not mention Har Megiddo (chapter 3, verses 14–16): *Multitudes, multitudes in the valley of decision, for the day of the Lord is near in the valley of decision. The sun and the moon shall be darkened, and the stars shall withdraw their shining. The Lord also shall roar out of Zion, and utter his voice from Jerusalem; and the heavens and the earth shall shake: but the Lord will be the hope of his people, and the strength of the children of Israel.*

We conclude that the Dead Sea transform has had an enormous effect on history. We cannot distinguish, however, some of the geologic disasters from those caused by people.

### San Andreas Fault and the 1906 Earthquake in San Francisco

The interior of North America is moving southward relative to the Pacific plate. That movement occurs along several faults in California (Figure 2.20), with the total displacement equal to the sum of the movements on the individual faults at different times. The total is controversial but amounts to several hundred kilometers over the past few tens of millions of years. The longest fault, and the one with the most displacement, is the San Andreas. Movement is almost continuous, with slippage and accompanying earthquakes varying from time to time on different segments.

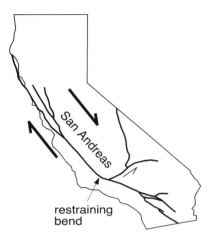

**Figure 2.20** San Andreas fault.

The most famous earthquake occurred at 5:12 AM on April 18, 1906, when the northern segment of the San Andreas slipped about 7 meters. The resultant earthquake had a magnitude of about 8 and an intensity estimated as 7 in San Francisco. Ground shaking lasted nearly 1 minute, and the few people who were awake at that time described streets that buckled up and down as waves passed.

The shaking caused many of the buildings in San Francisco to crumble, but the greatest damage resulted from breakage of water mains and gas lines. Some of the escaping gas caught fire, which ignited the wood that was used for most of San Francisco's buildings. The fire spread almost unchecked because of the absence of water to put it out.

The total carnage was colossal. In a city with a population of 400,000, about 3,000 people died and almost everyone was homeless. More than 25,000 buildings were destroyed in an area of 10 square kilometers. Most of the survivors had to be evacuated by a combination of ferries across San Francisco Bay and the Southern Pacific Railroad to the south. The U.S. government built 20,000 "relief houses" for people who stayed near the city.

Another result of the San Francisco earthquake is that people learned something about building in earthquake-prone areas. They realized that most of the destruction was caused by fire because of the rupture of gas and water lines and worked on better ways to build the lines. People also discovered that wooden buildings that didn't burn survived much better than buildings made of brick or adobe, which crumbled during the shaking. When tall buildings began to be constructed later in the 20th century, further research showed that buildings that sway during an earthquake suffer less damage than rigid buildings.

## Tsunamis

Tsunami is a Japanese word that means "harbor wave." It is an enormous system of waves in ocean basins that is caused by a sudden displacement of the ocean surface. This displacement creates a series of waves that move outward from the source at speeds up to 800 kilometers per hour. The wave length (horizontal distance between crests or troughs) is a few hundred kilometers, and the amplitude (vertical distance between crest and trough) is only a few meters. The amplitude is so low that ships at sea do not notice tsunamis, but the amplitude increases as the waves approach coastlines and "pile up."

Most tsunamis are caused by underwater earthquakes, and we discuss the tsunami in the Indian Ocean as an example. They can also be caused by massive landslides, and we describe a potential tsunami that could be created in La Palma as an example.

### Indian Ocean Tsunami

Normal life in the Indonesian province of Aceh ended abruptly at 7:58 AM (local time) on December 26, 2004. At this time the friction that prevented the Indian Ocean lithosphere from subducting beneath Indonesia was suddenly overwhelmed, and the lithosphere began to move downward. This motion created an earthquake whose epicenter was about 250 kilometers seaward from the capital city of Banda Aceh (Figure 2.21). Rupturing along the subduction zone then spread about 1,200 kilometers northward, causing earthquakes that lasted about 10 minutes. The strongest earthquake was generated at a depth of about 7 kilometers and had a magnitude slightly greater than 9. By the time the earthquake ended, the Indian Ocean lithosphere had moved more than 3 meters downward under the overlying crust. This movement also forced the seafloor above the subduction zone to rise more than 2 meters and tilted the ocean bottom near the coast of Sumatra downward.

A shallow earthquake of magnitude 9 would have caused enormous damage by itself, but the tsunami generated by movement of the ocean floor created even more destruction. Water receded from the Aceh shore a few minutes after the earthquake and then returned in a series of waves up to 30 meters high that raced inland for more than 1 kilometer. More

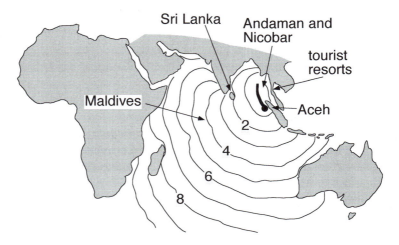

**Figure 2.21** Front of tsunami at 1-hour intervals across the Indian Ocean.

than 125,000 people died during that first set of surges. A more exact number is impossible to determine because many people or their bodies couldn't be found, particularly if they were carried out to sea by receding water.

As the tsunami destroyed the Aceh coast, the front of the wave pulse was already racing across the Indian Ocean at about 800 kilometers per hour (Figure 2.21). It reached the tourist areas on the west coast of Malaysia and Thailand about 1 hour after the earthquakes. More than 5,000 people died in Thailand, but the number of casualties was much smaller in Malaysia, which was protected by northern Sumatra. Before the tsunami reached the Thailand coast it killed some 5,000 people as it swept across the Andaman and Nicobar Islands, a chain of mostly volcanic islands owned by India.

The tsunami pulse reached Sri Lanka and southern India (Tamil Nadu) 2 hours after the earthquake and rushed ashore just as it did in Sumatra. More than 30,000 people died along the completely unprotected coast of Sri Lanka. The death toll in India, however, was only about 10,000 because the southeastern coast was partly protected by Sri Lanka.

Farther westward the tsunami arrived at the Maldive Islands about 3 hours after the earthquake. This former British colony became the Republic of Maldives in 1968 and has an elected government. It has a population of about 300,000, spread over more than 1,000, mostly uninhabited, islands that consist of coral reefs on atolls (see Hawaii). The islands have an average elevation of 2 meters, and only 28 of them have a surface area greater than 1 square kilometer. A tsunami surge nearly 5 meters high completely washed over most of the islands. About 80 people were confirmed dead, not because their bodies were recovered but because they couldn't be found.

The tsunami arrived at the Somali coast 7 hours after the earthquake and farther south in Africa a few hours later. By this time the surge was only about 1 meter high, and the steepness of the coastline prevented water from washing very far inland. Only a few deaths were reported on the African coast.

The total of more than 200,000 deaths caused directly by the tsunami was only part of the disaster. The long-term effects were still being felt more than 2 years later because:

- Buildings, roads, and harbors were completely demolished in some areas, particularly Sumatra, Thailand, Sri Lanka, and the Maldives. The lack of roads and ports made reconstruction difficult even for people and organizations that had money to pay for it.
- Utilities were similarly destroyed, leaving people without electricity and fresh water.
- The absence of utilities, particularly pure water, and the filth left in the demolished areas led to disease. The most serious were cholera, dysentery, and typhoid, and sick people could not be treated because of the destruction of medical facilities in the affected areas.
- The fishing industry was almost completely destroyed because fishing boats were demolished when they were dropped on land up to 1 kilometer inland from their harbors.
- Tourism became impossible when resorts were destroyed. This loss was particularly severe for the coast of Thailand and more so for the Maldives, which raised about one-third of its gross national product from tourist spending.
- Agricultural land inundated by seawater became nearly useless because of poisoning of the soil by salt.

Some governments and nongovernmental organizations (NGOs) offered immediate help. The U.S. government sent an aircraft carrier and supporting ships; they used helicopters to bring medicine and food to the stricken people. Australia also sent ships and relief supplies, and their government pledged 1 billion U.S. dollars of aid (about 2% of Australia's gross national product). Other countries and NGOs that offered close to 1 billion U.S. dollars include Norway, Croatia, and the World Bank. Former presidents George H.W. Bush and Bill Clinton toured the disaster sites and led an effort to raise relief money through private donations.

Resort owners encouraged tourists to return as soon as resorts could be repaired or even totally rebuilt. Within 2 years after the tsunami, new resorts in the Maldives were charging more than $1,000 per night for lodging. Some poor people who had nothing but a house and a fishing boat had to sell their land to anyone who had money, commonly large companies or people who lived outside the devastated areas. Some poor people even sold one of their kidneys to get enough money to live on. The economic disruption was still evident as long as 2 years after the tsunami, and the responsibility of governments for helping poor people was an unresolved issue.

Another question was whether better preparation could have reduced the effects of the tsunami. This preparation would have consisted both of early warnings to areas before the tsunami arrived and of better management of coastal areas.

The Indian Ocean did not contain a tsunami early warning system of the type that has been distributed throughout the Pacific Ocean. This system consists of seafloor sensors that detect slight pressure differences as tsunami waves pass over them and transmit this information to warn governments around the ocean of the approach of a tsunami.

After the tsunami, the government of Japan offered to fund the deployment of a warning system in the Indian Ocean, but the effectiveness would have been different in different areas. The tsunami arrived on the Sumatra coast so quickly that no warning could have been effective. The 1 hour between the earthquake and arrival of the tsunami at the coast of Thailand might have allowed more people to escape to high ground, and certainly a 2-hour warning to Sri Lanka could have saved many lives. No warning, however, could have had any effect on the Maldives, where there was no high ground to escape to.

Regardless of the warning system, however, better management of coastlines would have greatly reduced damage from the tsunami. All wave activity on shorelines is less effective if the coasts are protected by natural features such as coral reefs and mangrove swamps. Around much of the Indian Ocean, however, coral reefs had been destroyed to developed oceanic "farms" for shrimp and fish, and new ship channels had been cut through reefs. Coastal mangrove swamps had also been replaced in many areas in order to provide better access from the coast to the sea for fishing or the development of tourist resorts.

*Possible Tsunami from La Palma*

Cumbre Viejo volcano on La Palma (in the Canary Islands) is so steep that it is subject to frequent land sliding, and it shows the scars of numerous past landslides. Spanish geologists who study the volcano estimate that a mass of nearly 500 cubic kilometers is unstable on the western side of the volcano and highly likely to collapse into the Atlantic Ocean. If that happens, the tsunami generated could cause waves up to 10 meters high on the coast of western Europe and 20 meters in Florida.

## Living in Rifts

People live in rifts because they are low. These valleys contain lakes and rivers that provide water and fish. Their soils are more fertile than those of the surrounding uplands. Their low elevation also keeps them slightly warmer than the uplands. We discuss the East African rift system and Iceland here, and only mention two other rifts (Rhine and Rio Grande) that have been important in human history.

The Rhine River flows through a rift along most of its course between France and Germany. It was occupied by early people as they moved northward at the end of the last Ice Age (see Chapter 1). Romans built bridges across the Rhine and retreated across it to the west bank as their empire dwindled. The Germanic tribe of Franks built a fort at Frankfurt, just east of the Rhine. Much of the industry of modern Germany is in cities along the Rhine.

The second rift that we discuss only briefly is along the Rio Grande River in New Mexico. The Rio Grande valley has been occupied by Native Americans for more than 10,000 years (see Peopling of the Americas in Chapter 1). Spanish settlers established Albuquerque in 1706 after most of the native people retreated from the valley because of disease brought by Spanish conquistadors. Today most of the major cities of New Mexico are along the Rio Grande.

*East African Rift System (Great African Rift)*

The East African rift system has attracted people for more than 3 million years. One of the oldest known human ancestors is a partial skeleton found in the Afar area of Ethiopia in 1974 (Figure 2.5). She was an adult named Lucy, was 3.5 feet tall, and was classified as an *Australopithecine*. Olduvai gorge, in northern Tanzania, contains human fossils that range from about 2 to 1 million years old. Some anthropologists suggest that the human race evolved in the East African rift valleys.

The Afar region is still occupied. Most of it is the country of Djibouti. It has a nice seaport at the capital (also known as Djibouti), but the rest of the country shows how people are affected by living on a hot spot. Steam rises from the ground at several places, lakes are salty and hot, and vegetation is sparse. Pastoralists who live there are able to raise a few goats for milk and meat, but most of the food for Djibouti must be imported.

The East African rift has remained important in much of historical time. The fertility of the rift valley was a welcome change from the semiarid conditions on the sides, both for indigenous people and for European colonists. Lakes in the rift have always provided fish (see "Nile" in Chapter 1).

The rift valley was a principal target of the Imperial British East Africa Company, which was chartered in 1890 to produce economic benefits (make money) through trade and exploration in the Kenya region of East Africa. Operating from ports on the Indian Ocean, the company gradually expanded into the interior as far as modern Uganda. By 1902, colonists from Britain began moving into the area, and Kenya became a British colony. British power in East Africa was expanded after World War I, when the neighboring German territory of Tanganyika was ceded to Britain. Kenya became an independent country in 1963, and Tanzania (the former Tanganyika) in 1961.

### Iceland

If you like hot water, and plenty of it, you should live in Iceland. The intersection of two spreading centers with a hot spot/plume (Figure 2.14) provides both heat and water, and the abundant rain and snow yield even more water. These contributions partly overcome Iceland's isolation and far northerly position, where the northern fringe of the island is above the Arctic Circle. We explain the problems and benefits here.

The Icelandic rift that ends at Thingvellir (Figure 2.14) starts at the tip of the Reykjanes peninsula about 100 kilometers southwest of Reykjavik. The view along the peninsula is depressing—flat land where black basalt is mostly covered by various types of moss and lichen. Before the first people arrived in Iceland, the peninsula was a forest of birch trees, which used to cover nearly one-third of the country. They are gone now, however, because settlers began cutting them for timber as soon as they arrived. Because new trees cannot take root naturally in Iceland's poor soil and cold climate, now almost the only trees in Iceland have been deliberately planted in reforestation projects.

Reykjavik is a modern city of more than 150,000 people, about half of Iceland's total population. In addition to the typical attractions of any city, it has several indoor pools where Icelanders can frolic in geothermal water. The water that fills these pools also heats all of Reykjavik through a series of underground pipes that reach into every building and keep sidewalks in the center of the city warm enough to prevent them from icing during the winter.

Reykjavik means "smoky harbor" in the old Norse that is still the Icelandic language. It was settled as the Vikings began to move into Iceland during the 9th century. They remained around the edges of the island since the interior was, and still is, uninhabitable. Many of the first settlers did not come directly from the Viking homelands but indirectly from Celtic areas that the Vikings had subjugated. From these areas, particularly Dublin, the Vikings brought Celtic slaves, and this Celtic stock led to such racial mixing that modern Icelanders contain many more people with dark hair than their counterparts in Norway.

The rift at Thingvellir was the site of the world's first parliament, named the Althing. It was established in 930 AD in order to form a government and write laws that reflected the will of the inhabitants (women and slaves were not allowed to vote). A short distance from Thingvellir are Geysir and Gulfoss, the two most important tourist attractions in Iceland. Geysir gives its name to all of the world's geysers. Unfortunately it is no longer active, but its neighbor, Strokkur, erupts regularly. Gullfoss is the enormous Golden Falls, made by water flowing out of Iceland's interior.

Iceland's underground thermal water makes life on this isolated Arctic island much easier than it would be without the water. Pipes bring heat to all buildings at virtually no cost. Cheap electricity is produced by a combination of thermal springs and surface water. Heated greenhouses let Icelandic people raise tomatoes, peppers, and cucumbers to supplement sheep and fish, the only other food sources that are available locally.

The potential for producing more electricity is so large that the Icelandic government and private industry are planning the production of aluminum. Separating aluminum metal from its ore uses so much energy that the only cost-effective source of electricity is hydropower (see Chapter 4). If Iceland's full potential for electrical power is developed, some aluminum plants elsewhere in the world may be closed as companies ship their ore directly to Iceland. Some long-range plans envision laying an underwater power cable from Iceland to Europe so that electricity can be sold in the European markets.

**Living on Small Islands**

The world's oceans contain thousands of small islands. Some are the tips of volcanoes. Some are coral atolls that grew upward as their underlying volcanoes subsided. Only a few consist of continental rocks, such as the Seychelles, which were fragments of crust left in the Indian Ocean as India and Africa separated during the break up of Pangea (see Chapter 3).

Most islands are so isolated that they have virtually no effect on the world's economy except for tourism. A few, however, occupy strategically important positions and have been major contributors to world history. We describe Hawaii and Midway.

*Hawaii, Midway, and Pacific Airline Routes*

Before they became an American possession in 1898 and the State of Hawaii in 1959, the islands had a history that began at least 1,000 years ago. Polynesians coming from islands to the south navigated their outrigger canoes by the stars to bring their families across hundreds of miles of empty ocean. The canoes were large enough to bring pigs and other animals to islands where there had never been native land animals. The original Polynesian society consisted of small clans or tribes, each with its own ruler and its own small tract of land to control. In 1810 King Kamehameha brought all of these tribes under his control and ruled all of Hawaii.

The first missionaries came in 1820, and shortly afterwards American companies began to establish pineapple and coffee plantations. The owners of these plantations imported workers and new animals and plants that gradually replaced the native birds and plants. The immigrants also brought diseases that decimated the native Hawaiians, and the combination of deaths among the Hawaiians and immigration gives Hawaii a present population that is only about 10 percent Polynesian descent.

Midway atoll consists of a circular reef about 8 kilometers in diameter enclosing a lagoon with two small islands. Before discovery, the islands were barren of vegetation, except for some low shrubs, and were the home of enormous numbers of birds, which fed on fish attracted to the reef. Then, as now, the predominant bird is a variety of albatross known as "gooney birds" because of their clumsiness on land. Early American residents imported grass, trees, shrubs, and some topsoil in an effort to make it a more pleasant place to live and reduce the amount of sand hurled around the islands by wind.

**Figure 2.22** Airplane routes across the Pacific Ocean before World War II.

Unlike Hawaii, Midway was never inhabited by Polynesians, and it remained undiscovered until 1859. It was named Midway because the island lies very close to the 180° meridian, halfway around the world from Greenwich. The United States claimed Midway as a territorial possession in 1867, although the passage through the reef was so difficult that only small boats could reach the islands inside the atoll. In 1906 Midway became a way station for a trans-Pacific telegraph cable. In 1935, when Pan-American Airways developed a route between the United States and Manila, planes could not fly nonstop across the Pacific, and the air route made intermediate stops at Hawaii, Midway, Wake Island, and Guam (Figure 2.22). In 1938 Americans dredged a channel through the reef that allowed large ships to reach the islands and establish a naval air station.

Midway is a quiet place now, serving only as a Coast Guard station and emergency landing strip for trans-Pacific flights. In June 1942, however, Midway was the site of one of the most decisive battles of World War II.

The Japanese attack on Pearl Harbor, Hawaii, on December 7, 1941, was intended not only to destroy as much of the American navy as possible but also to cut air routes between the United States and the Philippines. For that reason, the Japanese also attacked Wake Island and Guam a few hours after the attack on Pearl Harbor. Wake Island held out until December 23, but Guam surrendered on December 10. With Wake and Guam securely in their hands, the Japanese armed forces waited until June to attack the better defended island of Midway. By that time, the American navy was ready for them and made the Battle of Midway one of the early turning points in the war in the Pacific.

The Japanese navy and troop transports advanced toward Midway in early June, 1942. Planes from the fleet attacked Midway on June 4 in order to prepare for a troop landing. Planes from Midway and the U.S. fleet attacked the Japanese fleet and troop transports immediately, beginning a battle in which no opposing ships ever got close enough to see each other. The battle continued for 3 days, resulting in the sinking of all six Japanese carriers and the loss of one American carrier. With all of their losses, the Japanese realized that landing on Midway would be impossible and sailed home on June 7.

# 3

# EVOLUTION, CREATIONISM, AND THE LONG HISTORY OF THE EARTH

The origin of living organisms is one of the most contentious issues facing people today. The two major opponents are those who believe in creationism and those who believe in evolution. Most creationists accept the biblical account in Genesis 1 and believe that all animals and plants were created in their present form during a 6-day period when God created the earth. Evolutionists believe that modern organisms evolved over a long period of time from older forms of life.

Arguments about creationism versus evolution are inevitably tied to questions about the history of the earth. Although some creationists accept an old age of the earth, young-earth creationists believe that the earth was formed in 4004 BC, an age calculated in 1654 by Archbishop Ussher. Evolutionists find that evolution is consistent with radiometric calculations that show that the earth formed about 4.5 billion years ago.

This section explores the evidence behind the controversies about evolution, creationism, and the history of the earth. We begin by showing how Archbishop Ussher calculated an age of 4004 BC for the earth. This calculation is followed by a discussion of the concept of "uniformitarianism" and the development of geologic knowledge of the long history of the earth.

The growing body of evidence for a long history of the earth was recognized by geologists during the 18th and 19th centuries. Charles Darwin knew about this evidence when he rejected creationism and proposed evolution. We discuss his evidence for evolution and then describe some of the violent arguments between people on different sides of the issue, concentrating on the present situation in America.

Darwin's ideas came from studies of living organisms, not fossils, but we show how the paleontologic record demonstrates the gradual evolution of life. The discussion concentrates on the first occurrences of new types of organisms and also on extinction events, when many old organisms died and were replaced by new ones.

The next parts of this chapter are discussions of the two methods for establishing geologic time. Combining the sequence of events recorded by rocks with the paleontologic record yields only a sequence of organisms and accompanying events. Geologists refer to the sequence as a "relative" time scale. An "absolute" time scale based on measured ages requires radiometric dating, and we discuss the process briefly.

Because no oceanic lithosphere is older than about 200 Ma, the long history of the earth is shown by the history of continents and supercontinents. Supercontinents are assemblages of nearly all of the earth's continental crust into one coherent landmass, and geologists have recognized at least four periods of assembly followed by four periods of break up. We finish this chapter by discussing these cycles.

## CREATIONISM

Creationists believe that Genesis 1 is an accurate account of the appearance of animals and plants in their present form during the 6-day period when God created the earth. Most creationists also believe that Genesis 1 requires a young age of the earth. Until the late 1800s, the age of the earth was calculated mostly by biblical scholars. The most famous was published in 1654 by Archbishop James Ussher, the Anglican Prelate of Ireland. He decided that God made the earth in 4004 BC. Ussher's calculation was consistent with those of several other scholars, all of whom had decided that the earth was no more than 6,000 years old.

Biblical calculations of the age of the earth are based on two assumptions. One is that the 5 days of creation specified in Genesis 1 before the creation of Adam and Eve were 24-hour days. The second assumption is that the ages of patriarchs at the birth of their oldest sons are accurately recorded in the Bible. For example, Genesis 5 states that Adam was 130 years old when he fathered ("begat") his son Seth, who was 105 years old when he fathered Enos.

Continuing this process through Genesis let biblical scholars calculate 1656 years between the creation of Adam and Eve and the construction of the ark by Noah just before the biblical flood (see Black Sea in Chapter 1). Later parts of Genesis and early chapters of Exodus showed that Moses led the flight from Egypt 2513 years after the creation of Adam and Eve.

The chronology of patriarchs after Moses is contained mostly in 1 and 2 Kings, with some in 1 and 2 Chronicles. 1 Kings 6 states that Solomon began to build the temple in Jerusalem 480 years after the Exodus from

Egypt, which would be 2,993 years after Adam and Eve. The patriarchs who followed Solomon in Jerusalem are listed in 2 Kings. They reigned for a total of 425 years until 3,418 years after the creation. At that time an army sent by Nebuchadnezzar of Babylon conquered Jerusalem, burned the temple, and took the Israelites to captivity in Babylon (2 Kings 25). There are no useful biblical records following the start of the captivity, but it is dated by a combination of Babylonian, Persian, and Greek records to have occurred at 586 BC.

Adding 586 to 3,418 gave Ussher and other biblical scholars a date of 4004 BC. for the creation. During the next two centuries most people agreed with this date or a close approximation.

Some problems with an age of 4004 BC quickly became apparent. One was (and still is) the significance of the 5 days of creation that Genesis 1 says precede the simultaneous creation of Adam and Eve. One uncertainty comes from Genesis 2, which states that all creation took place in 1 day, with Eve coming somewhat later than Adam because she was formed from one of Adam's ribs. Even if there were 5 days of creation before Adam and Eve, their significance is uncertain because the Bible makes it clear that a day is not necessarily 24 hours. For example, 2 Peter 3:8 states: *But do not forget this one thing, dear friends: With the Lord a day is like a thousand years, and a thousand years are like a day.*

For these reasons, many people feel that the concept of evolution is fully compatible with the biblical story of creation and a long age of the earth. Some people also accept the idea that God deliberately influenced evolution, and the Christian writer C.S. Lewis, among others, proposed that evolution was God's mechanism of creation.

## UNIFORMITARIANISM

Uniformitarianism is the principle that geologic processes now occurring on the earth are similar to those that occurred in the past. The concept developed gradually in the latter part of the 18th century and the early part of the 19th century. It is based on very simple observations, including: geologists can see different types of volcanic rocks forming now and assume that similar rocks that they find in studies of older rocks were also erupted from volcanoes; old sediments that look like modern beach sands were presumably deposited on ancient beaches; geologists can see that faults offset rocks when they cause earthquakes and can recognize faults that were active in the past. When Scottish geologist Charles Lyell began to publish his most famous book, *Principles of Geology*, in 1830, he described the basic principle of uniformitarianism as "The present is the key to the past."

The principle of uniformitariansm influenced James Hutton, a Scottish geologist who lived in the latter part of the 18th century before Darwin was born. (Most of the science of geology was founded by Scottish scientists, including Hutton, Lyell, and John Playfair, discussed in Chapter 1.) Hutton managed his family farms and realized that erosion of the soil would wear the earth's land surface down to sea level unless there was some process that formed new land. He regarded erosion and construction as one of the earth's natural "cycles." He also thought that these cycles operated continually, and when asked about the age of the earth, replied that he saw "no vestige of a beginning and no prospect of an end."

## EVOLUTION

Scholars throughout history have proposed various methods for the origin of animals and plants. Conflict between creationism and evolution began in the 18th century and intensified in the early 1800s. French anatomist George Cuvier found fossils of organisms that had become extinct and proposed that the earth had undergone several episodes of extinction and creation of new organisms. By this time scholars had translated enough Egyptian and Mesopotamian records to know that some civilizations were already in existence before 4004 BC. That observation led many people to think that the chronology in the Bible referred only to the Hebrews and not to all of civilization.

Despite these challenges, the biblical idea that all animals and plants were specially created at the same time dominated most scholarship in the 18th and 19th centuries. This Bible-based concept was most famously challenged in 1859 when Charles Darwin published his concept of evolution.

Darwin's opportunity to study animals and plants arose after the end of the Napoleonic wars in 1815, when Britain was clearly the world's foremost power. Because of its possessions all over the world, people could justifiably say, "The sun never sets on the British Empire." Accordingly, the British government commissioned the *Beagle*, a wooden ship powered by sails, to carry out surveys of the world's oceans and adjoining lands. The first voyage was not productive. On the second voyage, however, the captain asked Charles Darwin, then a student clergyman, to come along as a naturalist. The *Beagle* left Britain on December 21, 1831, and did not return until October 2, 1836. During this period of nearly 5 years, the *Beagle* circumnavigated the world and provided Darwin with an extraordinary amount of information on animals and plants.

Darwin's principal observation was how well animals and plants adapted to their environments. Even small differences existed among organisms that lived in slightly different environments. One observation, out of hundreds, was that finches in different parts of the Galapagos Islands had slightly different beaks that enabled them to eat different kinds of food.

Darwin decided that these variations showed that organisms survived if they were adapted to their environment and perished if they weren't. This belief led to a proposed mechanism for evolution—natural selection—which Darwin described in his most famous book: *On The Origin of Species by Means of Natural Selection; or, The Preservation of Favoured Races in the Struggle for Life.*

Darwin never felt that his concept of evolution was incompatible with a religious faith in God. He waited until 1859 to publish his work on natural selection, however, because he knew it would cause a furor by challenging the religious concept of special creation of each animal and plant. People who believed in creationism were particularly disturbed by the suggestion that humans were not specially created by God but were descended from other primates. They called Darwin "ape man" when he said that humans are most closely related to chimpanzees.

Now most people who think scientifically about the earth have come to agree with Darwin that all organisms on the earth have evolved by some process of natural selection. Part of their reason is that the paleontologic record can be interpreted as a consequence of evolution, and we discuss it after a summary of the present conflict between people who believe in evolution and those who believe in creationism.

## CONFLICT BETWEEN CREATIONISM AND EVOLUTION IN THE 20TH AND 21ST CENTURIES

Controversies about evolution and creationism continued vigorously through the 20th century, although mainstream Christianity had already accepted evolution as a divine mechanism of creation.

The most famous conflict between creationists and evolutionists in America was the trial of John Scopes in Dayton, Tennessee, in 1925. Scopes was the football coach at the local high school and occasionally filled in for teachers when they were absent. One day he was asked to handle a biology class that was about a section on evolution in the textbook. That class put him in violation of a Tennessee law that made the teaching of evolution illegal.

When Scopes was arrested, but not jailed, for the crime, everybody assumed that he would pay a small fine and the issue would end there. It didn't. The American Civil Liberties Union (ACLU) wanted to use the case to contest the Tennessee law. They alerted newspaper and radio reporters about the arrest, and stirred up defenders of evolution to rush to Scopes' aid. Then people opposed to Darwin came to help the prosecution. Two of America's most famous lawyers arrived to argue opposite sides of the case. They were followed by about 200 reporters, some of whom set up extra telegraph lines to get the news out quickly. Radio stations broadcast the proceedings. The courtroom wasn't large enough to hold all of the spectators, and the trial spilled onto the lawn.

The trial ended with Scopes being fined $100, which the lead prosecutor offered to pay. The high school wanted to hire Scopes for the next school year, but he declined in order to go to graduate school and later to work for oil companies.

The conflicts that have continued since the Scopes trial have been less famous but, in some ways, more virulent. One of the extreme positions is taken by "biblical inerrantists," who believe that every word in the Bible is true and God created every organism separately. The other extreme position is taken by people who reject the concept of a supreme being and believe all happenings in the universe are controlled by physical processes.

Many people occupy positions between these extremes. Some are paleontologists and other geologists who find that the process of evolution does not interfere with a belief in God. Some people take an apparently middle position and propose the concept of "Intelligent Design," which accepts evolution over a long period of time but insists that the process is so complex that it must have been guided by an "Intelligent Being" that most such believers regard as God.

Much of the controversy today is being contested both legally and politically. Science classes in public high schools in America generally use biology textbooks that discuss evolution but do not mention creationism. Consequently, creationists petition the elected school boards that control these schools to require different texts or add a supplemental text that describes creationism. Biology and other teachers generally believe in evolution and resist the change. Some biologists suggest that creationism should be taught in an interdisciplinary class or a class on religion, but not in science. Decisions are based on the beliefs of the members of the school boards, and the losing side commonly asks the court system to intervene and give orders to the school boards.

## PALEONTOLOGIC RECORD

Darwin's concept of evolution was based on living organisms instead of on fossils. It is supported, however, by the sequence of fossils that geologists have been able to recognize. We discuss the sequence here after clarifying some important issues about the classification of organisms.

When we look around us, it is easy to think that all organisms are either in the "kingdom" of plants or the "kingdom" of animals. We recognize plants by their ability to live by photosynthesis, which uses sunlight as their source of energy. We also understand that animals live either by eating plants, which are the "base of the food chain," or by eating other animals.

The actual classification of organisms is far more complex than the simple division between plants and animals. Another well recognized kingdom is fungi, which live by obtaining nourishment from the plant or animal hosts they live on, such as mushrooms on dead trees and athlete's foot on people.

More complexity is added to the classification by bacteria and other single-celled organisms. All bacteria are "prokaryotes," which means that their basic cell does not contain a nucleus, and the chromosomes are mostly attached to the inner wall of the cell. All bacteria live by eating other organisms except for cyanobacteria, which have some of the properties of bacteria but also photosynthesize and emit oxygen like algae and all other plants. That is why they are also referred to as "blue-green algae," and many modern varieties are called "pond scum" because they make messes by flourishing on the surfaces of still water.

We review the major episodes of evolution, and Figure 3.1 shows the places where some of the most important discoveries were made. This review is brief, and we refer readers who want more information to *Evolution 101* (R. Moore and J. Moore, Greenwood, 2006) and *Human Origins 101* (H. Dunsworth, Greenwood, 2007).

### Cyanobacteria

The oldest known fossils are cyanobacteria (blue-green algae). They occur in the Warrawoona Group, a suite of sediments with interbedded basalt in the outback of northwestern Australia. The sediments are chert, and radiometric dating of the basalts shows that the Group is about 3.5 billion years old. The Warrawoona sediments show several types of evidence of fossilized cyanobacteria. Some tiny filaments in the chert may

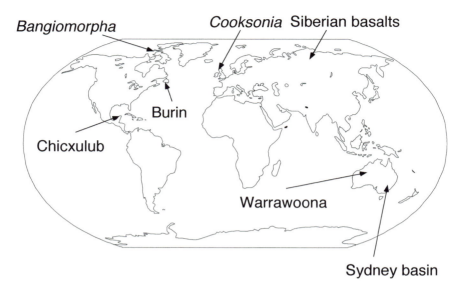

**Figure 3.1** Places discussed in the summary of evolution. Genus names are in italics.

be fossil algae themselves. Some structures seem to be "stromatolites." Stromatolites are layers of sediment that accumulate because cyanobacteria form sticky mats that trap particles of sediment as waves or other water movement wash the particles over the mats.

Cyanobacteria played an important role in the history of the earth because of their ability to pump oxygen into the atmosphere. Over time, this supply of oxygen was probably the main reason that the initial "reducing" (oxygen-free) atmosphere was converted into an "oxidizing" atmosphere that contained oxygen. The conversion is called the "Great Oxidation Event."

The exact time of transition is uncertain, but we can infer that it occurred about 2 billion years ago because of the change in iron-bearing sediments that were deposited before and after that time. Iron-bearing sediments deposited before about 2 Ga are "iron formations" and consist of chert with interbedded iron oxides (see "Iron industry" in Chapter 4). Iron-bearing sediments deposited after 2 Ga are "red beds" similar to the red sediments that now cover much of the earth (see Chapter 1).

**Algae**

An earth occupied only by prokaryotes came to an end during the Great Oxidation Event. The presence of oxygen in the atmosphere and oceans allowed the development of eukaryotes, which have their DNA in cell nuclei and which require oxygen in order to live.

No indisputable individual eukaryote cells have been found. Eukaryotic cells, however, unite to form multicellular organisms, called metazoans if they are animals and metaphytes if they are plants. Thus the oldest multicellular fossil is proof of the earliest evolution of eukaryotes.

At present the age of the oldest known multicellular fossil is uncertain. Some paleontologists have proposed that filamentous algae are preserved in the Gunflint chert, which is 1.8 billion years old and occurs with some of the iron formations in northern Minnesota. Other paleontologists have suggested that the oldest true multicellular organisms are preserved on Somerset Island, a bleak place of mostly bare rock in the far north of the Canadian Arctic. Some of the rocks on Somerset Island are sediments deposited at about 1200 Ma. The sediments contain tiny fossils of a red algae named *Bangiomorpha* pubescens.

Red algae are a type of marine algae that have a red color because they contain a pigment that absorbs blue light. This absorption leaves the light that is reflected from the algae mostly toward the red end of the spectrum. Red algae are one of the major types of algae, which, grouped together, are referred to as "seaweed." Japanese menus refer to red algae as "nori" and use it as a wrap for sushi. Some varieties of red algae are known as "coralline algae" because they secrete calcium carbonate and are an important part of reefs.

Whether *Bangiomorpha* pubescens evolved to other types of organisms is unclear, but its presence 1.2 billion years ago demonstrates at least the beginning of the evolutionary process that led to modern plants and animals.

## Oldest Animals

Newfoundland consists of several sets of rock suites (known as "terranes") that were swept together when North America and Europe collided during the assembly of Pangea. The eastern part of the island is the Avalon terrane, consisting mostly of fine-grained sandstones and siltstones. It is one of a series of terranes ("Avalonian") that were originally part of western South America and then moved across an intervening ocean to collide with eastern North America.

The Avalon terrane consists of peninsulas bordered by steep sea cliffs that reveal a remarkable set of fossils that span the Precambrian-Cambrian boundary. The oldest fossils are of organisms that lived some 10 to 20 million years before the development of animals that evolved into the types we know today. These oldest animals consisted only of soft bodies, and because they had no skeletal parts, they are preserved merely as impressions on the sedimentary layers. These organisms are called "Ediacaran" because most of them are almost identical to soft-bodied fossils originally found at Ediacara, Australia, and they are believed to be about 550 million years old. Because no Ediacaran fossils have ever been found in Cambrian or younger sediments, the Precambrian-Cambrian boundary is regarded as an "extinction event."

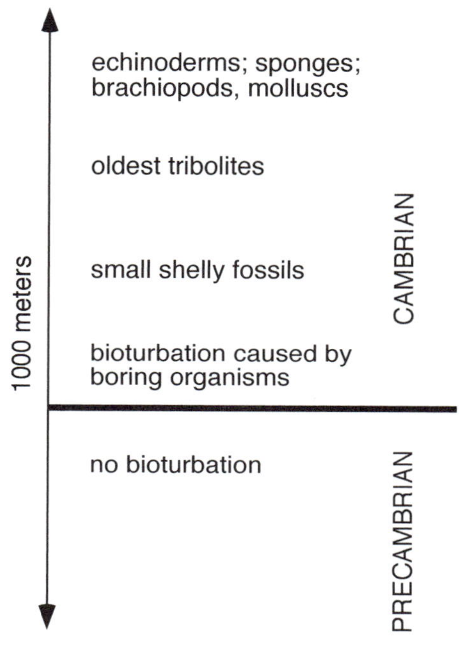

The most spectacular transition from Precambrian to Cambrian is displayed on the Burin peninsula, in the southwestern part of the Avalon terrane (Figure 3.2). Here the steep sea cliffs expose a thick sequence of slightly tilted sandstones and siltstones that were deposited over a period of several million years, during a time from approximately 540 to 530 million years ago. The lowest (oldest) sediments contain no fossils that anyone has been able to find. They are all finely layered, and the bedding planes are absolutely flat because the sediment was never disturbed by burrowing animals (the sediment is not bioturbated).

**Figure 3.2** Fossil succession at Burin peninsula, Newfoundland.

The flat layering is penetrated by tubes and pods of unlayered sand about 100 meters above the lowest exposed sediments on the Burin peninsula. The tubes are probably filling burrows made by worms that scavenged for food in the newly deposited sediment (the sediment is bioturbated). Slightly higher in the section are horizons of small shells that are commonly a few millimeters in diameter and made of calcite. They are simply called "small shelly fossils" because they are not like the shells of any younger organisms. Their exact classification has never been determined, and some fragments may simply have been small calcareous plates randomly stuck to the bodies of much larger nonskeletal animals to provide some protection against predators.

About 100 meters above the small shelly fossils are the oldest (lowest in the section) fossils of recognizable animals. They are trilobites, a type of arthropod (segmented animals like modern lobsters), and they are the oldest organisms that are known to have been encased in hard skeletons. Trilobite skeletons are made of calcite, but by the time they evolved, other types of organisms were already secreting both calcite and silica.

Dating of rocks at the Burin Peninsula shows that the transition from a truly ancient world to one very similar to the present required no more than a few million years. The reason for this rapid transition is unclear, but it may have been caused by increase in the concentration of oxygen in the atmosphere. The enzymes that control secretion of hard parts ("biomineralization") in animals require oxygen in order to function properly, and it is possible that the development of animals in the early Cambrian occurred because atmospheric oxygen concentration increased past some critical threshold.

## Oldest Vertebrates

All of the earliest animals were invertebrate, and animals with spines did not evolve until late in the Cambrian. The first vertebrates were fish. They appear to have developed in both fresh water and ocean water at the same time, suggesting that the first fish originated in coastal swamps and then diversified to other habitats. The oldest fish were jawless (agnathid), and the only agnathids existing today are minor groups such as lampreys.

These jawless fish were the root of the evolutionary tree that diversified to modern fish (with jaws), amphibians, reptiles, and mammals.

## Oldest Land Plant

Vascular plants include all plants except moss and algae. Their vascular system includes two different types of veins (conduits) to transport fluids throughout plants. Xylem carries water from roots through plants, and in trees it is the major component of wood. Phloem distributes food made in plant leaves to all parts of the plant.

The oldest fossils of vascular plants are in sediments of Late Silurian age in County Tipperary, Ireland. They are fragments of a genus known as *Cooksonia*, which is found as more complete fossils in slightly younger sediments in Wales. Although it is only about 2 centimeters high, Cooksonia had all of the features required by vascular plants. It had roots that connected the vascular system with the ground, veins that

distributed water and nutrients, and sporangia (spore cases) that prop-
agated the next generation of plants.

It seems almost impossible, but all of our modern trees, bushes, and
flowers ultimately descended from this tiny little plant.

**Permian-Triassic Extinction**

About 250 million years ago, a sudden event caused the extinction
of more types of organisms than had perished at the end of the Cretaceous. This extinction separates the Permian period of the Paleozoic
era from the Triassic period of the Mesozoic era, and it is particularly well displayed in sediments near Sydney, Australia. The Sydney
basin contains a series of sandstone and mud that were deposited both
before and after this extinction and, consequently, provides an excellent opportunity to discover the reason for the Permo-Triassic extinction.

Work by John Veevers of Macquarie University, in Sydney, Australia,
and Greg Retallack, now at the University of Oregon, showed that Permian sediments in the Sydney basin contain an abundance of coal
formed from decayed vegetation. The lower part of the Triassic, however, is just sediments without coal. This "coal gap" extends into the
later Triassic, when decaying vegetation began to accumulate again to
form more coal. Permian plants included *Glossopteris* and other varieties
that lived in cold regions around the edges of Permian glaciers (see
Gondwana) plus enormous cone-bearing trees (gymnosperms) with flat
leaves. Late Triassic plants included conifers and other varieties more
similar to modern trees.

The coal gap occurs not only in the Sydney basin but also at numerous
localities around the world, and many different explanations for it have
been suggested. Naturally, geologists searched for evidence of a meteor
or comet impact. They found none, nor did they find a tectonic event
that could have caused extinction by rearranging the patterns of ocean
and atmosphere circulation. Then in 1992, Ian Campbell and colleagues
at Australian National University, in Canberra, proposed the possibility
of alteration of the atmosphere and oceans by an enormous series of
basalt eruptions in northern Siberia (Figure 2.12).

These eruptions coincided with the Permo-Triassic extinction and
could have caused it in two ways. One method would have been to put
enough sulfur dioxide in the atmosphere to cause worldwide acid rain.
This rain would have poisoned both land plants and the oceans where
marine plants lived. Then since plants are the base of the food chain

for all animals, most animals would have died when the plants became extinct.

A second method by which the Siberian basalts could have affected the atmosphere is by emission of massive volumes of carbon dioxide that might have caused catastrophic global warming. A very hot atmosphere would have destroyed most existing land plants and, thereby, the animals that fed on them. High temperatures in the oceans would have reduced the solubility of oxygen, thus killing most marine animals.

Apparently it took several million years before new varieties of plants were able to establish themselves in sufficient abundance to form more coal and to provide an adequate food base for new types of animals to flourish in the later Triassic.

## Oldest Mammal

The principal differences between modern mammals and reptiles are that mammals have hair and nurse their young through a long period of development. Reptiles, however, have scales and lay eggs that hatch young reptiles that have almost the same range of abilities as an adult. Early mammals, however, may have lacked hair and laid eggs. Consequently, paleontologists commonly distinguish early mammals from reptiles by the anatomical feature that mammals have only one bone in their lower jaw. This single bone gives mammals a stronger lower jaw than that of reptiles, and mammals can shred food, whereas reptiles can only chop it.

Evolution of some types of reptiles to mammals probably occurred in the early to middle Mesozoic (Triassic and Jurassic). The first mammals were small, mouse-like creatures that were obviously dominated by larger and more numerous reptiles.

Mammals did not become dominant until after the Cretaceous-Tertiary extinction destroyed most of the earth's reptiles.

## Cretaceous-Tertiary Extinction

Geologists have long pondered a mystery recognized by the earliest people to study the earth. Fossils, both animals and plants, at the top of the Cretaceous period of geologic time are almost completely different from those of the overlying Paleocene, the lowest part of the Tertiary period. The most famous of the changes is extinction of the dinosaurs, but all types of animals and plants were affected, both on land and sea. All marine and flying reptiles became extinct, and the only surviving reptiles were crocodiles, snakes, and small lizards. In the oceans, several types of reef-forming organisms and cephalopods (ammonites and belemnites) also became extinct. Some paleontologists estimate that nearly 90 percent of all Cretaceous types of organisms (genera) became extinct, with the early Tertiary plants and animals gradually evolving new forms from the few genera that survived.

In some places Cretaceous (symbol K) and Tertiary (symbol T) rocks are separated by only a few centimeters of sediment commonly known as "boundary clay." The question is how such an enormous paleontologic change could occur so abruptly at this "K/T boundary." The answer apparently can be found within the boundary clay, which was first discovered near the town of Gubbio, Italy, by Walter Alvarez of the University of California, Berkeley.

Boundary clays have now been located at numerous places around the world. All of the clays are rich in iridium, an element very rare on the earth's surface. Iridium, however, is abundant in meteorites. The clays also contain small spherules of glass formed by melting of terrestrial rocks. Presence of these materials reinforced proposals that the K/T extinction had been caused by the impact of an enormous asteroid. The problem was where?

Then geologists working around the Caribbean noticed that tsunamis had eroded and/or deposited enormous volumes of sediment at the end of the Cretaceous. This suggested an impact near the coast of the Caribbean, and geologists began to look at the rock from a core drilled for oil exploration near Chicxulub, Mexico. It turned out to be melted terrestrial rock with an age of 65 million years, the exact age of the K/T boundary. The crater formed by the impact is buried by younger sediments, but geophysical studies show that is more than 150 kilometers across and was probably was formed by an asteroid about 40 kilometers in diameter.

The Chicxulub impact generated both immediate and long-term effects. The immediate result was to vaporize and melt both the asteroid and the rocks it plowed into. This put a fiery rain of particles over much of the earth, causing widespread fires. The fires explain the observation that the first plants of the Tertiary were mostly ferns, which have been observed to be the first plants to grow in areas scorched by modern forest fires.

After the fires, the next effect would have been global cooling caused by all of the ash left in the atmosphere. The freezing was then followed by heating caused by carbon dioxide from the limestones that the meteorite hit. This global heating would have lasted perhaps several thousand years until the oceans and newly developing plants could remove enough carbon dioxide from the atmosphere.

## Pleistocene Megafauna Extinction

Many large animals (megafauna) became extinct in the past 50,000 years. They died before and after the end of the last glaciation (see Chapter 1), and some scientists have attributed their extinction to their inability to adapt to rapidly changing environments. Other scientists, however, propose that the megafauna became extinct because people killed them. We explore this controversy here.

Mammoths (wooly mammoths) roamed Arctic and temperate regions during the last Ice Age and were a major food source for people. Most of the mammoths died when glaciation came to an end about 15,000 years ago, and whether they died because of hunting or because of climate change is unclear.

Many other animals in North and South America died at the same time as the mammoths or slightly earlier. They include mastodons, saber tooth tigers, sloths that lived on the ground, and horses, which were reintroduced to the Americas by European colonists (see Chapter 4).

Similar extinctions occurred in Australia about 25,000 years ago and New Zealand about 1,000 years ago. One of the victims in Australia was Diprotodon, which was 10 feet long and the largest marsupial that ever lived. A principal casualty in New Zealand was the Moa, a flightless bird that weighed up to 500 pounds.

Extinction of megafauna in the Americas began slightly before people arrived and was completed after people crossed the Bering Strait about 13,000 years ago (see Peopling of the Americas in Chapter 1). Extinctions in Australia occurred shortly after Aborigines arrived and in New Zealand soon after Maoris arrived.

The dodo, a flightless bird about 1 meter tall, was alive on the island of Mauritius when Europeans arrived in the 17th century. Shortly after their arrival, it was "dead as a dodo."

This coincidence between extinction and the first arrival of people supports a proposal by Paul Martin of the University of Arizona that the death of the megafauna by overhunting was a "blitzkrieg."

## RELATIVE TIME AND THE GEOLOGIC TIME SCALE

Based on the principles of uniformitarianism, geologists have long been interested in the sequence of events that affected the earth. They use very simple methods to determine this sequence. One is the "law

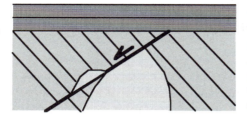

**Figure 3.3** Example of use of geologic relationships to determine relative ages of events. This diagram shows the following sequence: (1) deposition of a sequence of sediments (sand pattern); (2) tilting of the sediments by folding; (3) intrusion of a granite; (4) offset of both sediments and granite by a normal fault; (5) erosion to form an unconformity; (6) deposition of limestones (rectangle pattern) on the unconformity.

of superposition," which states that sediments are deposited on top of older rocks and become younger upward in the sedimentary section. This principle was used by William ("Strata") Smith, a British geologist who drew the first geologic map in 1799. It was near the city of Bath, England, and Smith expanded his work to a map of the entire British Isles in 1815.

Relative sequences can be inferred from other geologic features, and we use Figure 3.3 to explain them. It shows that:

- Rocks are older than igneous rocks that intrude them.
- Sediments derived by the erosion of rocks are younger than the rocks that were eroded.
- Faults are younger than rocks that they offset.
- Deformed (folded and tilted) rocks may be eroded to flat surfaces before younger sediments are deposited on top of them. These surfaces are called "unconformities" (see Chapter 2), and they may also be deformed later.

In the 19th century, geologists using these relationships recognized that different types of fossils occurred in rocks of different ages. Then they used these age relationships and the fossils in the rocks to construct an early time scale. The oldest time scales were divided into four parts, from youngest at the top to oldest at the bottom:

- Quaternary—Very young rocks with fossils almost identical to modern organisms.
- Tertiary—Rocks that contained fossils similar to modern organisms.
- Secondary— Rocks that contained fossils very different from modern organisms.
- Primary—Very old rocks that did not contain fossils.

This simple time scale was almost completely abandoned by the latter part of the 19th century. Primary and secondary were eliminated, and a new set of terms was used for the major eras of geologic time:

- Cenozoic—A combination of two Greek words: *kainos* = recent, and *zoic* = life. The Cenozoic includes the Quaternary and Tertiary.
- Mesozoic—Medium life.
- Paleozoic—Old life
- Proterozoic—Beginning of life.
- Archeozoic (now referred to Archean)—Before life.

Geologists also were able to subdivide the youngest three eras into periods (Figure 3.4). These periods were named for places where the rocks were best exposed and/or first investigated. The oldest rocks of the Paleozoic were first studied in Wales, and geologists recognized three suites of different ages. The oldest is Cambrian, named after Cwmbria, the Welsh name for Wales. The next youngest are Ordovician and Silurian, both named after old Welsh tribal groups.

Cambrian, Ordovician, and Silurian rocks in Wales are all deformed and metamorphosed, but they are overlain by a set of flat-lying, undeformed sandstones that lie on an unconformity over the older rocks. These sediments were named the Old Red sandstone, and the period was named Devonian because the rocks were first studied in Devon, England. The Old Red sandstone is widespread in the British Isles, and the concept of unconformity was first recognized at the base of the Old Red sandstone at Siccar Point, east of Edinburgh, Scotland. It is also an excellent building stone, and the castle in Inverness and St. Magnus cathedral in the Orkney Islands are both built of Old Red Sandstone.

The period of time younger than the Devonian was named Carboniferous because of the abundance of coal (carbon) in it. North American geologists subdivided it into a lower unit named Mississippian for outcrops in Mississippi and an upper unit named Pennsylvanian for rocks in Pennsylvania.

Rocks that were deposited in the youngest part of Paleozoic time are not as widely distributed as older Paleozoic rocks or rocks deposited in the Mesozoic or Cenozoic. The best uppermost Paleozoic outcrops are west of the Ural Mountains, and they are named Permian after the nearby city of Perm, Russia.

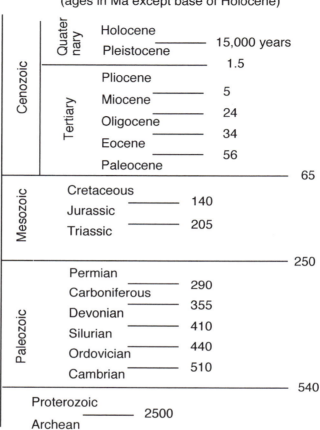

**Figure 3.4** Geologic time scale. The ages are radiometrically determined.

Names of the Mesozoic periods are derived from outcrops in Europe. Triassic was selected as a name for a series of rocks that was subdivided into three well-defined units. The name Jurassic comes from the Jura Mountains of southern France. Cretaceous rocks contain an abundance of chalk (fine-grained limestone). The chalks are very widely distributed, including the White Cliffs of Dover, and the name Cretaceous comes from kreide, the German word for chalk.

The Cenozoic era includes the Tertiary and Quaternary periods. They are further divided into epochs, largely based on the sequence of sediments around Paris, France. The epochs in this Paris basin are all named with prefixes and "cene" as a suffix:

Paleocene—from the Greek *paleo*, meaning old;

Eocene—from the Greek *eos*, meaning dawn;

Oligocene—from the Greek *oligos*, meaning few;

Miocene—from the Greek *meion*, meaning less;

Pliocene—from the Greek *pleion*, meaning more;

Pleistocene, from the Greek *pleistos*, meaning much;

Holocene—from the Greek *holos*, meaning whole.

The Quaternary period includes the Pleistocene, which is essentially the time of the Ice Ages, and the Holocene, which is the 15,000 years since the end of the last Ice Age (see Chapter 1).

## RADIOMETRIC DATING AND THE ABSOLUTE TIME SCALE

Paleontologic, stratigraphic, and other geologic information yield only a relative time scale. Determination of absolute ages requires measurement of radioactive decay systems. We explain this process in three parts: radiometric dating of very old events using heavy isotopes; use of the K-Ar system to date events that are particularly important for archaeologist and anthropologists, and $^{14}C$ (carbon-14; radiocarbon) dating of events in the past several thousand years. We follow these explanations with some of the absolute ages that geologists have discovered.

### Radiometric Dating Using Heavy Isotopes

Because the rate of decay of a radioactive isotope is constant, the number of atoms that decay at a time is proportional to the amount of the isotope present. Thus, as decay continues, the number of atoms that decay at any time decreases as the amount of parent decreases. Half life is the time in which one half of an isotope decays. After one half life, only one half of the original isotope remains. Then one half of that remainder decays after one half life, and only one-fourth of the original remains. This process continues, with the amount of parent reduced by one half after each period of one half life.

In order to calculate absolute ages, we need to explain the process more quantitatively.

Let P be the radioactive parent isotope and D be the stable daughter isotope produced only by the decay of P. Let $P_t$ and $D_t$ be the amounts of parent and daughter in a rock that we measure today. Then the amount of P in the rock when it was initially formed is $P_0 = P_t + D_t$. The radioactive decay equation is:

$$P_t = P_0 e^{-\lambda t},$$

where $\lambda$ is the decay constant, $e$ is the base of natural logarithms, and $t$ is the time during which P has decayed. The equation can also be expressed as in $P_t/P_0 = -\lambda t$.

The relationship between $\lambda$ and half life is shown by the equation

$$t = \text{half life when } P_t/P_0 = P_t/(P_t + D_t) = \frac{1}{2}.$$

These equations let geologists calculate the ages of formation of rocks by measuring $P_t$ and $D_t$ in rocks and using known decay constants (half lives) of the isotopes.

Radiometric dating began in the middle of the 20th century with the use of uranium-lead systems and has now expanded to the use of several decay systems. The principal ones and their half lives are:

- $^{238}U$ yields $^{206}Pb$ with a half life of 4.47 billion years;
- $^{235}U$ yields $^{207}Pb$ with a half life of 0.704 billion years;
- $^{232}Th$ yields $^{208}Pb$ with a half life of 14 billion years;
- $^{87}Rb$ yields $^{86}Sr$, with a half life of 48.9 billion years;
- $^{147}Sm$ yields $^{143}Nd$ with a half life of 106 billion years.

The isotopic systems can be used to date either whole rocks or individual minerals in the rocks. One of the most common methods is to use U-Pb systems to date the mineral zircon ($ZrSiO_4$) (which can contain trace amounts of U) in granites, and many reported ages of granites are actually the ages of their zircons.

**Radiometric Dating Using the K-Ar System**

Natural potassium consists of 94 percent $^{39}K$, 6 percent $^{41}K$, and only about one part in one million $^{40}K$. The $^{40}K$ is unstable and decays both to $^{40}Ar$ and $^{40}Ca$ with a half life of 1.26 billion years. Argon can remain only in cold rocks because it is a gas. Consequently lavas and hot volcanic ash contain no Ar when they are erupted, and any Ar in them must have been produced by radioactive decay after the rocks cooled. Ages of eruption have been calculated from $^{40}K/^{40}Ar$ ratios for rocks as old as 1 billion years. Ages are most reliable, however, for rocks with ages from 100,000 to a few million years. This range is particularly useful for anthropologists, who can date sediments that contain human fossils by the ages of interbedded volcanic rocks.

**Radiometric Dating Using Carbon-14 (Radiocarbon Dating)**

The carbon in atmospheric carbon dioxide is 99 percent carbon-12 and 1 percent carbon-13. A tiny amount (about one-trillionth) of the carbon, however, is carbon-14. It is formed by reaction of cosmic rays with nitrogen-14 according to the equation

$$^{14}N + \text{cosmic ray yields } ^{14}C + \text{beta ray}$$

The $^{14}C$ is unstable and decays back to nitrogen plus hydrogen with a half life of 5,700 years.

Plants absorb $^{14}C$ during photosynthesis and pass it on to animals that eat the plants. This process causes the concentration of $^{14}C$ to remain unchanged in both plants and animals while they are alive. When they die, however, no more $^{14}C$ is taken in and the concentration decreases as the $^{14}C$ decays. For example, the concentration of $^{14}C$ in plant or animal remains is only half as much as in living organisms after 5,700 years (one half life).

Radiocarbon dating is done on several types of organic remains. They include partially decayed plant debris in peat bogs, animal bones, and charcoal in the fire pits of prehistoric humans. Dates are valid up to about 20,000 years, after which the amount of $^{14}C$ remaining in the dead organism is so small that accuracy is uncertain.

**Measured Ages**

Absolute ages can be measured only on rocks that formed by crystallization. They include igneous rocks formed from a liquid and metamorphic rocks developed by solid-state crystallization of new minerals in older rocks. This relationship makes it necessary to infer the ages of most geologic features from their relationships to igneous rocks. One example is inferring that a suite of sedimentary rocks is older than a granite that intrudes them. Similarly, a fault that cuts a granite must be younger than the granite.

Geologists have used radiometrically measured ages and these geologic relationships to place absolute ages on the time scale and to date events such as mountain building and rifting (see Chapter 2). We show the ages of eras, periods, and the epochs of the Cenozoic in Figure 3.4.

One very important age that cannot be measured directly is the age of formation of the earth. Any rocks or minerals that were present at that time were apparently melted or otherwise destroyed. The oldest known age is 4.4 Ga for a zircon in sedimentary rocks from Western Australia that were derived by the erosion of granites that were destroyed by later geologic processes. The age of 4.4 Ga was measured by Simon Wilde and colleagues at the University of Western Australia and proves that the earth formed before 4.4 Ga.

The value currently accepted for the age of the earth was determined from studies of lead isotopes in meteorites by Claire (Pat) Patterson at Caltech. Ages of 4.55 Ga have been found for all meteorites, which are apparently remnants of the formation of the solar system. An age of 4.55 Ga for the earth is consistent with data on the earth's lead isotopic system, which indicates that the earth formed at the same time as other parts of the solar system.

## CONTINENTS AND SUPERCONTINENTS

Continents preserve the long history of the earth, and we must study continents if we want to understand that history. We divide this discussion into two parts: the formation of continents, and the history of supercontinents, which were assemblages of nearly all of the world's continental crust into one landmass. The oldest continents apparently formed at approximately 3 Ga, and there have been at least four times when all continents were assembled together.

### Formation of Continents

Present continents consist of two suites of rocks. One is Precambrian igneous and metamorphic rocks that are exposed at the surface in continental areas referred to as "shields" (Figure 3.5) North America contains the Canadian shield. The shields in Europe are in the Baltic countries and Ukraine. Asia contains shields in Siberia and India. The two shields in South America are mostly in Brazil and Venezuela. Africa and the Middle East contain five small shields.

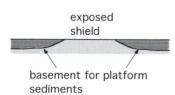

**Figure 3.5** Typical exposure of shield surrounded by sediments lying on the basement formed by buried margins of the shield.

Rocks exposed in shields extend outward from the shields and form "basements" for the deposition of sediments. These "platform" sediments lie unconformably on the basement, and most of them were deposited from shallow oceans that covered parts of the basement from time to time. The oldest platform sediments mark the time at which the continents became "stable" as they moved around the earth and were deformed only along narrow belts of mountain building or rifting. Most of the present continents became stable at different times in the Proterozoic.

The ability to study and date old platform sediments lets us recognize small continents that existed before modern continents were formed. Although the oldest zircon shows us that the oldest rocks existed 4.4 billion years ago (see Measured Ages), the oldest platform sediments did not develop until about 3 Ga. Several small continents developed at this time, and others formed later. The shields and covered basements of modern continents formed largely by the accretion of several of these small continents.

## History of Supercontinents

Supercontinents were assemblages of nearly all of the earth's continents into one landmass. They developed as continents moved around the earth, came together, and later dispersed. Their existence has always been contentious, and we tell the story of how different ones were proposed to exist, beginning with the first one to be named and proceeding to the most recent proposal.

### Gondwana

In the late 19th century, Austrian geologist Eduard Suess realized the similarity of fossils in Australia, India, and the southern parts of Africa and South America (Figure 3.6). He also noticed that all of these areas had undergone glaciation at some time in the past. In addition to the presence of rocks deposited by glaciers, Suess found that all four of those areas contained fossil leaves of a plant named *Glossopteris* that had never been found anywhere else in the world.

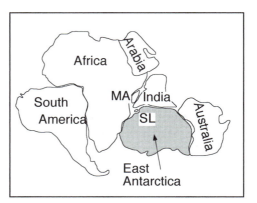

**Figure 3.6** Configuration of Gondwana. The position of East Antarctica was confirmed by *Glossopteris* fossils found on the Scott expedition.

Based on this information, Suess proposed that South America, Africa, India, and Australia had all been part of one landmass at some time in the past. He named it "Gondwana Land" (now Gondwana) after the Gonds, a group of people who lived in the hill country of central India. Suess had no way to determine the age of Gondwana, and because he also did not know about seafloor spreading, he thought that Gondwana had been a single landmass when parts of the Indian Ocean had been lifted up above sea level. Suess also didn't know whether Antarctica had been part of Gondwana, and that problem would not be resolved until after an expedition led to the Antarctic in 1910 by British explorer Robert Scott.

Dr. Edmund Wilson signed up to be the second in command on Scott's expedition. Wilson was officially a biologist, but he had such a wide-ranging interest in all of nature that he was also familiar with many aspects of geology. And when he wasn't studying biology or geology, or helping run the expedition, he made pencil and pastel sketches that still remain classics of Antarctic illustration.

Scott, Wilson, and their men sailed through the pack ice around Antarctica and arrived at Cape Evans, a rocky promontory of Ross Island. They built a hut that still stands, preserved as the survivors left it when rescue arrived in 1913. The hut is about 25 kilometers north of McMurdo Station, which is part of the U.S Antarctic Research Program.

After spending the winter at Cape Evans, Scott and his men started for the South Pole on November 1, 1911. They used ponies and dogs to haul sledges with supplies across the ice shelf that floats on the Ross Sea until they started to climb the Beardmore Glacier, one of several that come from the high Antarctic icecap and cut through the mountains down to the ice shelf. From there to the south the explorers man-hauled sledges up the glacier to a height of about 10,000 feet and then over the polar icecap. After sending supporting parties back north at various places, Scott, Wilson, and three other men arrived at the pole on January 17, 1912, only to find a tent and flag left by a Norwegian expedition led by Roald Amundsen and a note from Amundsen saying that he had arrived on December 16 (a full month earlier).

There was nothing for Scott to do except lead his men back over the same route they had taken south. Wilson's diary shows that he had time to collect some rocks along the side of the Beardmore Glacier as they went back down. He put them on their sledge to take back to Cape Evans.

One of the five men died on the glacier on the way north. Another one died after the group reached the ice shelf. Finally Scott, Wilson, and one other man reached a point 11 miles from a food depot they had left on their way south. They never made it. The last entry in Scott's diary is dated March 29, 1912. The last line is "It seems a pity, but I do not think I can write more."

Members of the expedition who spent a second winter at Cape Evans set out the next spring to find out what had happened to Scott and his men. On November 12 they found the tent with three bodies, diaries, and cameras with film waiting to be developed. The sledge next to the tent contained no food and 30 pounds of rock from the sides of the Beardmore Glacier. The rocks contained *Glossopteris* fossils.

At a cost of five lives, geologists all over the world would soon learn that Antarctica had been part of Gondwana in the late Paleozoic.

### Pangea

Pangea, also spelled Pangaea, means "all earth." It is the name given by Alfred Wegener to a supercontinent that used to contain almost all of the world's continental crust. When he first published his ideas in 1912, Wegener knew about Gondwana, and he decided that all of the other continents were once joined to Gondwana to form a supercontinent that later broke up into modern continents.

The evidence for the existence of Pangea has always been overwhelming. The easiest to see is how well modern continents fit together. The map shown here (Figure 3.7) is somewhat different from the one that Wegener drew, but it clearly demonstrates how modern continents can be organized into one continental mass. Actually, the fit is even better than Figure 3.7 shows. A projection of Pangea on a hemisphere, thereby showing many of the continents distorted, would show that North America and Siberia were in contact with each other, and the Arctic Ocean came into existence only when Pangea split apart.

At its peak of aggregation ("maximum packing") at about 250 Ma, Pangea was a linear block about 20,000 kilometers long from north to south and 8,000 kilometers wide from east to west. The southern part (modern Antarctica) was over the South Pole and the northern tip reached almost to the North Pole.

Soon after Wegener's proposal, paleontologists provided even more persuasive evidence for the existence and break up of Pangea. Plant and animal fossils with ages between about 250 Ma and 200 Ma are similar on all of the present continents. All variations between organisms of the same age could be explained by differences in latitude. Organisms that flourished in warm climates are found in continents that were near the equatorial center of Pangea. Those that grew in cold climates

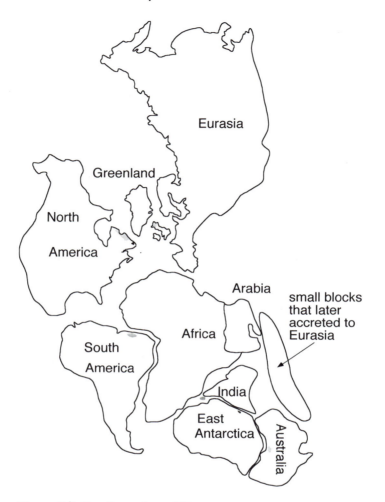

**Figure 3.7** Configuration of Pangea.

(including *Glossopteris*) occur in parts of continents that were near the poles.

This similarity among animals and plants continents began to disappear at about 200 Ma, and organisms that lived after that time developed very differently in different continents. A few examples of the enormous range of paleontological evidence that modern continents began to separate from Pangea starting at approximately 200 Ma include:

1. Monotremes (egg-laying mammals) evolved only in Australia and could not migrate to other continents.

2. Varieties of dinosaurs in Africa became different from those in South America at about 100 Ma.

3. Also starting at about 100 Ma, primates began to develop very differently in different continents (see Peopling of the Americas in Chapter 1). Apes evolved in Africa but not in South America. Monkeys with "prehensile" (grasping-type) tails evolved in South America but not in Africa.

With all of this evidence for Pangea, we might think that geologists around the world enthusiastically adopted the idea immediately. Wrong! Geologists in most of the world accepted it, but not those in the United States. They gave a large number of reasons for their disbelief. The continents didn't really fit, which was ridiculous. The paleontologists weren't any good, which was even more ridiculous.

Reading through some of the papers that these objectors wrote makes their problem clear. They were embarrassed because they hadn't thought of the idea and jealous of the person who had. They were doubly infuriated when they realized that Wegener was not a geologist but a meteorologist (Wegener died in 1930 attempting to bring supplies to a meteorological station in Greenland). American geologists were particularly violent in their attacks, possibly because Wegener was German, not American.

American geologists finally accepted the reality of continental drift 50 years after Wegener described it. This acceptance came when Harry Hess and Bob Dietz proposed seafloor spreading (see Chapter 2), which required that continents move around the earth.

### Rodinia

Mark and Dianna McMenamin were interested in the rapid evolution of life at the beginning of the Cambrian (see Paleontologic Record). In 1990 they published a book titled *The Emergence of Animals: The Cambrian Breakthrough*. They decided that this sudden evolution could not have happened unless a supercontinent had existed before the Cambrian, and they named the supercontinent Rodinia, from a Russian word that means "motherland."

McMenamin and McMenamin did not propose a configuration or specific age for Rodinia, but both configuration and age followed shortly. Three papers about Rodinia were published almost simultaneously in 1991 by Ian Dalziel, of the University of Texas, Paul Hoffman, now at Harvard, and Eldridge Moores, at the University of California, Davis. These authors noted the large number of orogenic belts with ages of about 1 Ga and proposed that Rodinia had been assembled at that time.

**Figure 3.8** Configuration of Rodinia.

The best configuration placed North America at the center of a globular supercontinent with the configuration shown in Figure 3.8.

### Columbia

The supercontinent Columbia was first proposed in an abstract published in 2000 by John Rogers of the University of North Carolina, Chapel Hill. The name comes from the Columbia region of the northwestern United States, which was thought to have been joined to eastern India in the middle of the Proterozoic (Figure 3.9). In 2002 Rogers and M. Santosh of Kochi University, Japan, published more complete information about Columbia and suggested an age of 1.6 Ga.

Guochun Zhao of Hong Kong University and three colleagues published a paper in 2002 proposing a supercontinent with an age of 1.8 Ga and a slightly different configuration from the one suggested for Columbia by Rogers and Santosh. Because of the slight difference in age and configuration, Zhao and his colleagues could have used a new name for their proposed supercontinent, but they graciously decided that a new name would cause confusion in the geological literature and decided to use their work to suggest a different configuration and age for Columbia.

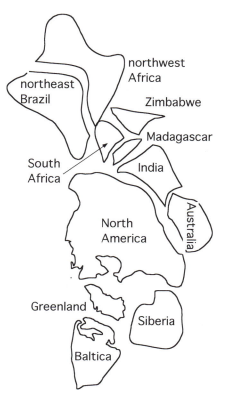

**Figure 3.9** Configuration of Columbia.

### The Supercontinent Cycle

The past existence of supercontinents requires mechanisms for both assembly of continents and their later breaking apart. Assembly presumably occurs where several subducting slabs come together and leave continents together on the surface as the slabs descend into the mantle. Kent Condie of New Mexico Tech has called these areas "slab avalanches." Break up presumably occurs because supercontinents prevent heat from escaping from the mantle until so much heat has built up that the mantle rises and cracks the supercontinent apart.

The complete cycle of accretion and dispersal seems to require several hundred million years. Possibly the next supercontinent will develop when the Pacific Ocean closes in a few hundred million years.

# 4

# RESOURCES AND THE ENVIRONMENT

Geology and atmospheric dynamics control the distribution of natural resources around the world. Reallocating these resources so that they are shared with those in countries other than where they originate has profoundly shaped world history, including power relationships, national boundaries, and the well-being of billions of people throughout geography and time.

We discuss five critical aspects of present and past consumption of resources and the environmental consequences of their use: (1) energy resources; (2) mineral and rock resources; (3) food resources; (4) fresh water resources; and (5) wireless communication and the Internet, which has recently become one of the world's most powerful resources.

Energy resources have been critical to cultural evolution throughout history because use of stored energy multiplies the productivity of human and animal energy. The activities involved in discovery, extraction, refining, purchase, delivery, and use of these energy resources constitute a significant proportion of world economic activity. All energy sources, however, create significant problems. The United States must import most of its oil, and world reserves are being depleted. Burning fossil fuels has serious consequences for global atmospheric processes (see Chapter 1). Nuclear fuels have several downsides, partly because radioactive elements can be used for making the most powerful weapons ever invented, and radioactive isotopes are very hard to get rid of safely once they have been created for any purpose. Fully renewable energy resources include hydropower, solar and wind energy, and biofuels, but production of some biofuels reduces the amount of food available for people.

Mineral and rock resources have been critical for making tools and other physical goods throughout history and also for serving as symbols of wealth. We begin by showing how the progressive discovery of new mineral and rock resources, including fertilizer, allowed significant expansion and development of existing cultures. We also discuss how the fierce desire to obtain wealth has driven both development and conflicts among world peoples. We explore the history and uses of gold and diamonds as examples.

Without food and water, all of the mineral and energy resources a country can obtain won't be worth much. We explore several aspects of food resources in this chapter, including people's nutritional needs, food available to people in different societies, and the dramatic food exchange between the Old and New Worlds when European explorers reached the Americas. We also describe the consequences of the wrong prediction of famine that Thomas Malthus thought would result from population growth.

Looking more broadly at the environment evokes questions of how humans, who are globally connected to an extent unprecedented in history, will manage to live successfully in the long term. These issues are often addressed under the heading of "sustainability," which means patterns of living and working that can continue, or be sustainable, for the foreseeable future and beyond, without serious damage to any member or component of ecosystems.

Issues of sustainability inevitably bring up questions of justice. It is becoming clear that the actions of people in both industrialized and developing nations have a significant unintended effect on poor people around the world. One example is that millions of people driving vehicles that emit large amounts of $CO_2$ into the atmosphere affect the livelihoods of people who live in coastal areas, particularly in island countries, and are threatened by rising sea level.

How can we learn to live better? That is, to live well and in ways that do less damage to other humans and other creatures within our ecosystems? Can you think of ways to do this in your own community?

## ENERGY RESOURCES

We begin this chapter by defining the units that are used to measure quantities of energy and power. This identification is followed by discussions of production and consumption in the United States, worldwide production and consumption, fossil fuels, nuclear energy production, nuclear hazards, and renewable energy.

Energy production and consumption are commonly measured in British thermal units (Btus). The principal unit for describing worldwide or national consumption is a "quad," which is equal to one quadrillion Btus.

Oil was originally shipped around the world in barrels containing 42 gallons, and it is still measured in barrels although all shipments now are in tankers that are simply large vats of oil.

Power is energy generated per unit of time. A typical unit of power is a kilowatt, which equals 0.95 Btu per second.

Other units of energy and power are summarized in a table in the Glossary.

All data about energy use are from the U.S. Energy Information Agency (www.eia.doe.gov).

**United States Energy Production and Consumption**

United States energy consumption generally increases from year to year. In 2006 the production (in quads) from various sources was as follows (addition of individual numbers does not equal the total because of rounding errors):

| | |
|---|---|
| Fossil fuel | |
| Coal | 22.511 |
| Natural gas | 22.931 |
| Oil | 39.758 |
| *Total fossil fuel* | 85.200 |
| Nuclear | 8.208 |
| Renewable energy | |
| Hydropower | 2.889 |
| Geothermal | 349 |
| Solar | 70 |
| Wind | 258 |
| Biomass | 3.277 |
| *Total renewable* | 6.832 |
| *Total (all sources)* | 100.24 |

The summary shows that Americans depend on fossil fuel for 85 percent of their energy. The only other significant source is nuclear power (see Nuclear Power). Although Americans are very enthusiastic about renewable energy, very little has actually been developed.

Different sources of energy are used in different ways. Most of the coal consumed in America is used in boilers that produce steam to generate electricity. Some power plants use natural gas and oil. The only significant electrical generation except from coal-fired plants, however, is from nuclear energy plus a small amount of hydropower. Almost all oil is used for vehicles. Natural gas and some oil are used for direct heating of buildings.

All of the energy sources in the United States are domestic except for oil. In 2006 the United States imported more than 60 percent of its petroleum, a percentage that has been increasing steadily since the 1970s.

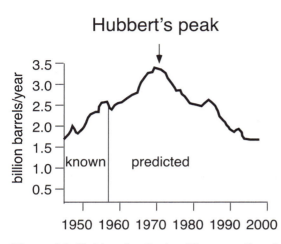

**Figure 4.1** Hubbert's Peak. The predicted curve was based on data for U.S. oil production up to 1956.

Conversion of the United States from oil exporter to oil importer had been predicted as early as 1956 by M. King Hubbert, a geologist who spent much of his career with Shell Oil Company in Houston, Texas. In 1956 he delivered a paper titled "Nuclear Energy and the Fossil Fuels" to a conference about the oil industry. In that paper he compiled enough information about past oil production and known reserves to predict future production in the United States. We show his prediction in a graph that has become widely known as "Hubbert's peak" (Figure 4.1).

Hubbert's conclusion was based on the realization that the area under the production curve must equal total production. The mathematical expression is

if $p(o)$ = annual oil production, then total production
$= \Sigma \ p(o)$ over all production years.

This conclusion showed that American oil production could not continue increasing as it had until 1956. The United States had produced a total of about 90 billion barrels of oil by 1956, and Hubbert estimated that future production would come from the remaining 150 million barrels of untapped reserve. Based on these numbers, Hubbert estimated that U.S. production would peak at about 3 billion barrels per year at some time around 1970.

Hubbert's prediction turned out to be correct. American oil production peaked at about 3 billion barrels/year in 1970 and has declined steadily since then. This decline was slowed, but not stopped, by discovery of reserves at Prudhoe Bay on the north slope of Alaska in 1967. It was similarly slowed by discovery of oil in deep waters seaward of the continental shelf in the Gulf of Mexico.

## World Energy Production and Consumption

Per capita energy consumption varies enormously from country to country. The United States, with 5 percent of the world's population, consumes 25 percent of world energy production. Other industrialized countries also have high per capita consumption. Poor countries, conversely, have very low rates of energy use and are forced to use hand labor for jobs that are routinely done by powered equipment in industrial countries.

In 2006 total energy consumption by the world was 457,609 quads, distributed as follows:

| | |
|---|---|
| Oil | 169,277 |
| Coal | 122,246 |
| Natural gas | 105,331 |
| Hydropower | 28,997 |
| Nuclear | 27,473 |
| Renewable | 4,285 |

About half of that energy consumption was in the 30 countries that belong to the Organization for Economic Development (OECD), which include: Australia; Austria; Belgium; Canada; Czech Republic; Denmark; Finland; France; Germany; Greece; Hungary; Iceland; Ireland; Italy; Japan; Luxembourg; Mexico; Netherlands; New Zealand; Norway; Poland; Portugal; Slovak Republic; South Korea; Spain; Sweden; Switzerland; Turkey; the United Kingdom; and the United States.

The other 162 countries that are members of the United Nations must survive on the half of world energy production that is not used by OECD countries. The extraordinary contrast between energy consumption in poor and rich countries is shown by data compiled by the World Resources Institute (www.wri.org). The Institute shows per capita consumption in 2006 in units of kilograms of oil equivalent (kgoe), which is the amount of energy generated by burning 1 kilogram of oil (approximately $\frac{1}{3}$ gallons).

The countries with the highest per capita kgoe are all major oil exporters, such as Qatar, Bahrain, the United Arab Emirates, Brunei, Canada, and Norway. The five highest per capita energy consumers that are not oil exporters are:

| | |
|---|---|
| Iceland | 11,718.0 kgoe |
| Luxembourg | 9,408.8 |
| United States | 7,294.8 |
| Sweden | 5.764.8 |
| Singapore | 5,158.7 |

The five lowest energy consumers are:

| | |
|---|---|
| Bangladesh | 160.9 kgoe |
| Eritrea | 199.3 |
| Senegal | 233.2 |
| Myanmar (Burma) | 276.5 |
| Yemen | 294.8 |

Total and per capita energy consumption continues to grow in most of the world. The most rapid growth is in China and India. They are the world's most populous countries and also among the most rapidly developing. Because most of their growth is by burning coal, they are regarded as major threats to increased emission of $CO_2$ and global warming (see Global Warming in Chapter 1).

**Fossil Fuel**

This discussion describes the formation of fossil fuels and then the history of their use.

*Formation*

Fossil fuels include coal, oil, and natural gas. They are all derived from the decay of plant and animal remains, and their burning releases solar energy that was originally stored in organisms by plant photosynthesis.

Photosynthesis is a catalytic process that uses the energy in sunlight to help plants synthesize carbohydrates. The process depends on chlorophyll, a complex molecule that occurs in green plants. Chlorophyll is able to absorb energy from the sun and release electrons. The electrons then cycle through a series of reactions that lead to carbohydrates before returning to the chlorophyll. The overall equation for the photosynthetic process is:

$$CO_2 + H_2O + energy \text{ yields carbohydrates} + O_2$$

Conversion of carbon dioxide and water to carbohydrates and oxygen has several major effects. It creates the atmospheric oxygen that all animals need to breathe. Furthermore, formation of carbohydrates provides the base of the food chain for all animal life. Removal of carbon dioxide from the atmosphere also has an important effect on the temperature of the atmosphere (see Global Warming in Chapter 1).

When plants and animals die, atmospheric oxygen begins to break their organic molecules down to $CO_2$ and $H_2O$. In some environments, however, the organic matter is preserved because oxygen cannot reach it. On land, this preservation occurs where plants die in swamps whose water contains only a small amount of oxygen. Similarly, plants and animals that die in the ocean fall into poorly oxygenated water and may be buried by younger sediments before the organic material is completely destroyed.

The conversion of recently deposited organic material to any type of fossil fuel requires burial. Burial increases both the heat and pressure and causes complex changes in the types of organic molecules preserved in the sediment.

Coal develops as heat and pressure increase on the peat that is originally formed in a swamp. Slight increase in pressure forms lignite, which is a poor fuel. Further increase in temperature and pressure forms bituminous (black) coal, which is the major coal mined worldwide. Higher temperatures and pressures produce anthracite, which is a hard black coal that consists almost entirely of carbon.

Oil and natural gas pass through a process known as "maturation." During this process, the originally preserved carbohydrates gradually lose all of their oxygen and are converted to various types of hydrocarbons. The lightest of these hydrocarbons is methane ($CH_4$), which is a gas and constitutes almost all of the material marketed as natural gas. Hydrocarbon molecules that contain large numbers of carbon atoms, such as octane ($C_8H_{18}$), are liquids, and oil contains many different kinds of these molecules.

Oil and gas must be recovered from drilled wells. Oil and gas float upward on water in the pores of rocks and accumulate in "reservoirs" that are overlain by impermeable rocks that prevent the oil and gas from rising any farther (Figure 4.2). Oil and gas in reservoirs commonly are at high pressure because they are trapped between water at the bottom of the reservoir and the overlying rock. Oil and gas in some reservoirs are at a pressure high enough that they flow out of wells naturally, but oil in most reservoirs has to be pumped to the surface.

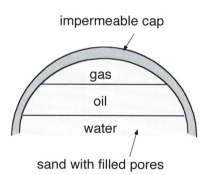

**Figure 4.2** Oil reservoir in anticline. Impermeable cap rock covers porous reservoir rock. Water fills all open spaces (pores) in the ground, and the oil reservoir forms where oil and gas float on the water but are trapped underground by an impermeable rock.

After mining or pumping, all fossil fuels release energy by the reverse of the photosynthetic process:

Fuel + $O_2$ yields $CO_2$ + $H_2O$ + energy

### History of Fossil Fuel Use

Wood and brush were the only significant fuels used throughout most of human history. Some of the wood was converted to charcoal by burning it in piles covered by dirt or other material that kept enough air out that the wood was not entirely consumed. In addition, some people burned cattle dung when there wasn't enough wood available, and people who lived in the far north occasionally burned blubber (fat) from seals and other animals. Candles also burn animal fat (tallow).

Very small amounts of oil were used as long as 2,000 years ago. Both Romans and Chinese scooped oil from seeps, where some oil had escaped from underground reservoirs and reached the earth's surface (see Silk Road). This oil was used partly in lamps to supplement the light provided by candles, which have been in use for at least 5,000 years.

Energy sources had become slightly more complex by the 1700s. Coal was the principal fuel for the industrial revolution. Stoves began to replace fireplaces, which waste much of the heat released by burning wood or coal. Whale oil, which was very expensive, was used in lamps. Methods were developed to make "coal gas" by combining coal and water, and this development led to networks of gaslights in cities by the early 1800s.

Enormous improvements in technology developed during the 1800s. The steam engine that had been invented by James Watt in the late 1700s became very widely used by the early part of the 1800s. Early steam engines produced up-and-down or side-to-side motion, but rotary engines were soon developed. Rotary engines operate mostly by boiling water and turning wheels by directing the steam at projections on the outside of the wheels.

Rotary movements were essential for the development of electrical generators by Michael Faraday in the 1830s. Faraday discovered that iron bars could be magnetized by passing electrical currents through wires wrapped around the bars. Similarly, magnets rotated inside a coil of wire "induced" electrical currents in the wire. These developments brought electricity into wide use, although the first effective electric light bulb was not developed until the 1880s by Thomas Edison.

Production and use of oil developed rapidly in the middle 1800s. The first modern oil well was drilled in 1848 at Baku in modern Azerbaijan. The process of making kerosene from oil was developed in 1849, thereby replacing whale oil in lamps with the much cheaper kerosene. A crew led by Edwin Drake drilled the first oil well in the United States (in western Pennsylvania in 1959), and America immediately became a major oil producer.

Oil supplies were very important during World War II. Oil produced in the United States was sent by tanker ships to all of the Allies. Although both Germany and Japan tried to use submarines to interrupt the flow, the Allies had plenty of oil throughout the war.

In contrast, Germany and Japan had very few oil supplies. One of Japan's first goals was to capture the oil fields of Indonesia, but they were able to use the oil only before the American navy secured dominion over the Pacific Ocean. One of the first German objectives in Russia was capture of the Baku oil fields, but their advance toward them was stopped at Stalingrad (now Volgograd).

The oil fields at Baku are still contentious in the 21st century. Fields even larger than those at Baku have been discovered throughout much of the Caspian Sea basin. Currently all of the oil is sent out by pipelines through Russia, but private companies and several nations are looking for a more reliable route (Figure 4.3).

The oil and gas industry changed greatly in the 20th century. Oil is moved all over the world by supertankers. Oil and gas pipelines cross entire

**Figure 4.3** Possible routes from the Caspian Sea to world oil markets.

continents and some lie on sea floors. Because gas is much more plentiful than oil, some gas is liquefied so that it can substitute for oil as liquid natural gas (LNG). The present marketplace is completely global as people compete for the cheapest source of energy.

**Nuclear Energy**

Naturally occurring uranium in the earth consists of two isotopes: 99.3 percent is $^{238}$U with a half life of 4.47 billion years; 0.7 percent is $^{235}$U with a half life of 0.704 billion years. The shorter half-life shows that $^{235}$U is much more radioactive than $^{238}$U. Both isotopes have long decay series that ultimately form different isotopes of lead. Intermediate members of each decay series emit either alpha (helium nucleus) or beta (electron) particles plus gamma radiation (energy) according to the overall equations:

$$^{238}\text{U} - 8\alpha \text{ yields } ^{206}\text{Pb} + \text{energy}$$

$$^{235}\text{U} - 7\alpha \text{ yields } ^{207}\text{Pb} + \text{energy}$$

In the 1930s, several physicists realized that uranium was sufficiently unstable and that it could undergo a different type of breakdown—fission. Fission is the splitting of a nucleus into two parts when the nucleus is made even more unstable than it usually is by penetration of neutrons into the nucleus.

Fission of an individual nucleus by one neutron produces one neutron that can penetrate a second nucleus. By about 1940, however, nuclear scientists realized that they might be able to design a "reactor" in which several nuclei would undergo fission at the same time and produce more neutrons than were needed for the initial fission. This process is referred to as a "chain reaction," and it can occur only when enough uranium is put into a small space to form a "critical mass." Fission releases energy because the total mass of the fission products is smaller than the mass of the original nucleus. The energy of the reaction is calculated by the Einstein equation $E = mc^2$, where $E$ is the energy released, m is the mass lost, and c is the speed of light.

Natural uranium does not contain enough $^{235}$U to be fissionable. The minimum concentration to create a chain reaction is about 4 percent $^{235}$U. This "enriched" uranium is formed by separating the $^{235}$U from natural uranium and adding it to other batches of natural uranium. The separation is very difficult because it depends solely on the difference in masses of the two uranium isotopes. The principal method is to convert the uranium to $UF_6$ (uranium hexafluoride), which is a gas and can be centrifuged to separate the heavier $^{238}UF_6$ from the slightly lighter $^{235}UF_6$.

Uranium enriched to several percent $^{235}$U is used in reactors. Atomic bombs can be made from uranium enriched to very high percentages of

$^{235}$U. The $^{238}$U that remains after extraction of $^{235}$U is designated "depleted uranium." Depleted uranium has such low natural radioactivity that it is safe to handle, and it is used for such purposes as metal casings for artillery shells.

The first chain reaction in a nuclear reactor occurred on December 2, 1942. It happened in a reactor built by Italian physicist Enrico Fermi in a squash court in the athletics area of the University of Chicago. The reactor contained enriched uranium and also rods consisting of stable elements that absorbed enough of the neutrons to keep the reactor from going out of control and, perhaps, generating so much heat that it would melt.

Construction of a reactor was the first part of an American effort known as the Manhattan Project. The purpose of this project was to design an atomic bomb before German scientists, who were known to be working on one of their own, could finish a bomb and possibly win the war. In order to explode, the atomic bomb would have to consist solely of highly enriched uranium without control rods, thus making it very different from reactors where fission was tightly regulated.

Soon after nuclear reactors were built, scientists recognized that not all of the uranium breaks apart by fission, but a small percentage of it simply adds neutrons or protons to the $^{238}$U nucleus and to form heavier elements, referred to as "transuranics." The most important of these elements found during the Manhattan Project is plutonium ($^{239}$Pu), which forms when a proton penetrates a $^{238}$U nucleus. Plutonium is as fissionable as $^{235}$U, and the atomic bomb dropped on Hiroshima was made of $^{235}$U, whereas the bomb dropped on Nagasaki was made of $^{239}$Pu.

Shortly after the end of World War II, scientists found a nuclear reaction more powerful than fission. Fusion occurs when light elements, particularly hydrogen, are forced to bind together to form elements that are slightly heavier. Hydrogen bombs are made by using uranium or plutonium bombs to compress hydrogen until it fuses and releases energy explosively.

In 1946 the United States established the Atomic Energy Commission, and in 1955 the United Nations developed the program, Peaceful Uses of Atomic Energy. Both organizations sought to divert some of the scientific efforts from building atomic weapons during the Cold War to finding industrial uses of nuclear technology.

One of the principal efforts to find peaceful uses for atomic energy was to build nuclear reactors to generate electricity. Most of these reactors used uranium enriched in $^{235}$U, but the possibility of using plutonium led to the development of reactors specifically designed to convert $^{238}$U to $^{239}$Pu. These "breeder" reactors essentially made it possible to use $^{238}$U in addition to $^{235}$U to generate energy. In all of the early reactors, the nuclear core was simply a source of heat to boil water and generate electricity by using the steam to turn turbines.

Rapid construction of nuclear reactors increased the need for uranium ores. Most uranium ores are in hydrothermal veins and consist of pitchblende, a black mineral with the approximate formula $U_3O_8$. When exposed to oxygen in air or surface water, pitchblende breaks down to form the highly soluble uranyl ion ($UO_2^{+2}$). This solubility causes uranium ions to be widely distributed in surface waters, where they precipitate in sediments as various minerals that commonly have bright yellow or green colors. Uranium also comes out of solutions that encounter organic matter, which reduces the uranyl ion to insoluble $U^{+4}$ oxide.

Shortly after it was founded, the U.S. Atomic Energy Commission developed a program to encourage people to locate uranium deposits. Many amateur prospectors came to the deserts of the Southwest because some ore deposits had already been found there. The town of Moab, Utah, was advertised as "The uranium capital of the world" because so many prospectors came to it.

People who searched for uranium could use two instruments to aid their search. One was a Geiger counter, which detected radiation and was named after the German scientist Hans Geiger (an ardent Nazi). The second tool was an ultraviolet ("black") light that caused the yellow and green uranium minerals to fluoresce. Many prospectors looked for ores at night because they could not see the fluorescence during the day.

During the first few years of the 21st century, world uranium mining is about 40,000 tons per year from all types of deposits. The principal exporting countries are Canada, Australia, and Kazakhstan, and the principal importing country is America. Uranium is traded mostly as a material called "yellow cake," which consists primarily of uranium oxide. Although virtually no new nuclear reactors are being built, most of the uranium is used to replace "spent" fuel that has been in a reactor so long that it is no longer fissionable (see Nuclear Hazards).

## Nuclear Hazards

Naturally radioactive isotopes are not very dangerous, but isotopes produced in nuclear reactors and weapons are extremely hazardous. These isotopes pose two problems. One is in the fallout from aboveground nuclear explosions, and the other is storage of waste from nuclear reactors. We define units used in measuring radiation first and then discuss problems caused by radiation.

- 1 roentgen of radioactivity produces 1 electrostatic unit of electricity in 1 cubic centimeter of air.
- 1 rem (radiation equivalent mammal) is the amount of radiation absorbed by a mammal exposed to 1 roentgen of radiation. The difference between roentgen and rem is that different organisms and different parts of animals vary in their ability to absorb radiation.

### Fallout from Bombs and Reactor Explosions

Fallout causes damage in several ways. Immediate radiation effects have been studied mostly at Hiroshima and Nagasaki. The radiation dose received by people was estimated by determining how far they were from the blasts. These data indicate that people exposed to more than 1,000 rem died almost instantly, but people who received a few hundred rem died later, largely because of cancer and destruction of white blood cells.

Two isotopes in fallout are particularly hazardous. One is $^{90}$Sr, which emits beta rays and has a half life of 28 years. The danger posed by strontium results from its ability to substitute for calcium in Ca-bearing compounds. Bones consist mostly of calcium hydroxy phosphate (the mineral hydroxyapatite), and $^{90}$Sr in their structure weakens them by bombardment with beta rays. The normal path for $^{90}$Sr to get into bones is for people to drink milk from cows that grazed in pastures contaminated by fallout, and one emergency response in areas of nuclear accidents is to bring food in from areas that have not been contaminated.

The second dangerous isotope is $^{131}$I, a beta emitter with a half life of 8 days. The half life is so short that the isotope wouldn't be dangerous except that iodine is necessary for functioning of the thyroid gland, which consequently absorbs iodine. Most isotopes of iodine are not radioactive, but any radioactive iodine that enters the thyroid can destroy it. The principal response to contamination by $^{131}$I is to give people potassium iodide pills, which saturate their thyroids with so much normal iodine that $^{131}$I cannot find a place in them.

### Nuclear Waste

Nuclear waste is nuclear fuel that must be removed from reactors when it has decayed so much that it no longer generates enough heat and must be replaced by new fuel. This spent fuel is very dangerous and has to be stored somewhere safe. The general intention is to isolate isotopes from ground water for 10 half lives, by which time only 0.1 percent of the original material remains ($2^{-10} = 0.001$)

Isolation is not difficult for fission products, none of which have half lives longer than a few tens of years. The problem is transuranics, most of which have very long half lives. In particular, $^{239}$Pu is relatively abundant and has a half life of 29,000 years.

The problem of long-term storage of nuclear waste is highly contentious. Some people think that they can shut down the nuclear industry if they can prevent construction of any place to store the waste. Others want to enjoy the benefits of nuclear power so long as the waste is stored NIMBY (not in my backyard!).

In the early 21st century, the major waste storage site considered by the U.S. government is at Yucca Mountain, Nevada. Yucca Mountain consists of a suite of volcanic and sedimentary rocks in the Nevada Test Site, where nuclear weapons have been tested for more than 50 years. Despite this location, many people have asked Congress and the court system to keep nuclear waste out of Yucca Mountain.

**Renewable Resources**

Energy consumption continues to rise as conventional resources dwindle. This problem has caused an intensified search for renewable resources in the 21st century. Renewable resources are ones that replenish themselves as they are used, and we discuss four possibilities here: solar power, water power, wind power, and biofuels.

*Solar Power*

The intensity of solar radiation is highest at the equator and diminishes toward higher latitudes. Averaged over the whole earth, however, each square kilometer receives about 0.03 quadrillion British thermal units (Btus) of energy every year. The total amount of energy from all sources used by all of the people in the world each year is about 450 quadrillion Btus (450 quads). Quick calculation suggests that the entire energy needs of the world could be satisfied by the sunlight that falls on only 15,000 square kilometers. The problem is to devise ways to use this energy effectively.

The process starts by converting sunlight into electricity in photovoltaic (solar) cells. These cells contain materials that are unstable when struck by solar radiation and lose electrons from their crystal structures. Coupling the unstable materials with an electrical conductor lets electricity flow out of the cell.

Generating electricity leaves several problems to be solved. We mention only a few:

- Should the electricity be generated in large solar farms and distributed from there, or should it be generated in small local grids?
- Should electricity from solar grids merely replace current electrical power, or should it be used for other purposes? Replacing gas and oil used in heating would leave those fuels to be used largely by the plastics industry.
- Should cars and trucks be run on batteries that are recharged at local power stations instead of running on gasoline that is bought from filling stations?
- Should solar electricity be used to hydrolyze water, thus generating hydrogen as a portable fuel?

What is the best way to store electricity so that it can be used at night or on cloudy days? Can we design batteries to store enough electricity to run power grids? Is it possible to build a worldwide power grid so that energy is always generated by solar panels somewhere?

Until these and similar questions are answered, solar power will probably not be a major part of the world's energy supply.

### Water Power

Water at a high elevation has more "potential energy" than water at a lower elevation, and the difference in water pressure between two elevations is regarded as the "hydraulic head." The energy that water loses when it runs downhill has been used for most of human history to turn waterwheels. Now, however, running water is used almost entirely to generate electricity.

Electricity is generated by running water through turbines. Turbines contain a series of blades set at an angle that makes the core of the turbine rotate as water passes through. As we discussed in the section "Fossil Fuel" earlier, this rotation of a magnet within a coil of wire generates electricity. In theory, turbines could simply be placed in a river, but it is far more efficient to put them in a dam so that they will continue to run as long as water remains in the lake behind the dam.

Dams are used for two purposes in addition to power generation. One is to ensure a steady supply of water to people who need it. This water supply is reliable because the lakes created by dams are much less likely to run dry than rivers. A second use of dams is flood control. Dams can reduce flooding below the dams because they can store water until it reaches the crest of the dam.

The various uses of dams frequently compete with each other. People who need water and electricity want the lakes behind dams to be kept full. Conversely, people who are concerned about floods want the lakes behind dams to be relatively empty so that they can store more water when necessary.

Dams have many important uses, but they can also be dangerous. Sudden collapse of a dam sends a rapid flood of water downstream that causes destruction of property and sometimes kills people. Dams also have profound effects on the environment. One problem is interruption of the movements of fish, including those that swim upstream from oceans in order to spawn (anadromous fish). A second problem is that artificial lakes may saturate rock in the walls of a valley and cause landsliding.

All of the uses and problems of dams cause difficulties when people want to construct dams to generate electrical power. No new dams have been built in the United States in the 21st century, and many existing ones are considered in danger of collapse.

### Wind Power

Wind generators look very much like propellers of airplanes. Windmills have been used for thousands of years, but modern ones are mostly used to generate electricity. They do so by rotating magnets in coils of wire in the same way that electricity is generated by all other forms of power. Some generators are as small as ordinary fans and are used to supply power only to a single building. Some generators, however, have blades several meters long, and many of them can be linked together to form enormous "wind farms."

No serious environmental effects have been attributed to wind generators, and their importance in power generation will probably increase in the future.

### Biofuels

Biofuels are plants or recently dead plants that are used to generate heat. They can be prepared in a large number of ways. One of the most important is to distill ethanol from corn or other plants. Ethanol is an alcohol, identical to the alcohol in alcoholic beverages. Ethanol also burns and can be used in automobile engines, possibly after slight engine modification. For this purpose, ethanol is mixed with regular gasoline to form a mixture that generally contains no more than 20 percent ethanol. In addition to slightly reducing the consumption of regular gasoline, ethanol, or other compounds that contain oxygen make gasoline burn more completely and reduce emission of unburned gases.

A type of biofuel that is becoming more important is "biodiesel," which its proponents hope will replace the petroleum-based diesel fuel now used in trucks. Biodiesel fuel is formed by reacting an alcohol (methanol or ethanol) with oils that have been squeezed out of plants. Plant oils are large fatty acids, and their reaction with alcohol yields diesel fuel plus glycerin by a process known as "transesterification."

Biofuels are produced both from edible and nonedible crops. Biodiesel is normally produced from plants that are not edible by humans or animals. Consequently, its production should have no effect on food supply. Ethanol, however, is produced from corn (maize) or other crops that are edible by both humans and animals. Although only a small amount of the corn crop in the United States is eaten directly by people, using corn for fuel would affect food supply because it is a principal food for cattle. Feeding corn to cattle in fattening pens makes their meat more tender when they are slaughtered, and removal of some of this corn would leave consumers either with less beef or with beef that is less tender than they are used to.

The use of edible crops to produce fuel instead of food has raised a great deal of concern. It centers on the question of whether it is ethically proper to reduce the world's food supply by using edible crops to make fuel. This question must be resolved before biofuels become a more important part of the world's energy supply.

## ROCK AND MINERAL RESOURCES

We obtain the raw materials for our societies from the earth, and our needs change continually through time. Ten thousand years ago, people looked only for rock. Now, however, we require an enormous variety of ores. This review subdivides this discussion into base metals, iron and iron alloys, nonmetals such as crushed rock, the history of use of these materials, fertilizer, and precious metals and jewels. All of the information about production of metals is data for 2006 from the U.S. Geological Survey (http://minerals.usgs.gov).

### Base Metals

Base metals are used for a wide variety of commercial purposes without being alloyed with iron to make steel. The principal ones are aluminum, copper, lead, zinc, and tin.

Aluminum is mined from tropical soil known as bauxite. It develops in areas covered by forests where water in the soil leaches out $SiO_2$ and leaves a residue of aluminum oxides/hydroxides. Consequently, the major producers of aluminum ore are tropical, including Guinea, Australia, Jamaica, and Brazil.

The aluminum–oxygen bond is very strong, and separating metallic aluminum requires a large supply of electrical energy. For this reason, countries that produce aluminum ore generally do not recover aluminum metal but ship the ore to processing plants that have adequate electrical power (commonly hydropower). These plants are mostly in industrialized countries, but they are becoming more widespread as countries with excess hydropower, such as Iceland, try to attract processing plants (see Chapter 2).

Aluminum is much lighter than other metals and does not corrode easily. The lightness makes it ideal for airplane bodies, and lack of corrosion makes aluminum the principal metal used in the manufacture of cans for food and beverages.

Copper occurs in a large number of minerals, mostly as sulfides or oxides. The most productive mines are open pits in large bodies of granite that have been thoroughly mineralized with copper ores. Major production is from the Andes, particularly in Chile and Peru, and the United States also produces substantial amounts of copper ore. Copper metal is relatively easy to recover by smelting, and most of it is used for electrical wiring.

Lead sulfide (galena; PbS) and zinc sulfide (sphalerite; ZnS) commonly occur together in veins. The metal–sulfide bonds are comparatively weak, and both lead and zinc can be smelted from the ores easily. The major producing countries for both metals are China, Australia, and Peru.

Lead is used mostly in batteries now. Before it was recognized as a health hazard, however, it was used for a variety of purposes. Lead tetraethyl ("ethyl") was added to gasoline as an "anti-knock" agent to make the gasoline burn more smoothly. The low melting point of lead made it an ideal solder, and it added luster to both glass and paint. All of those uses are now banned in the United States and most industrial nations. Lead in gasoline has been replaced by organic compounds that contain oxygen, and solder is made with tin plus small amounts of minerals such as silver and antimony.

Zinc is used largely for two purposes. It makes brass when combined with copper, and it reduces corrosion of steel when combined with iron to form galvanized steel.

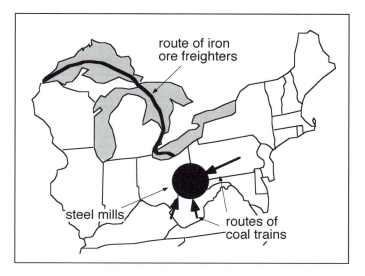

**Figure 4.4** Pittsburgh and nearby areas became centers of iron and steel production because iron ore could be shipped through the Great Lakes and coal could be sent by train.

Tin occurs in veins and placers of cassiterite ($SnO_2$). It is mined in Southeast Asia, Bolivia, and China. Tin is used in solder and as a plating on iron to make cans that are stronger than aluminum cans.

### Iron and Iron Alloy (Ferro-Alloy) Metals

Iron occurs in a wide variety of ores, the most important of which are "iron formations" (see Chapter 3). Iron formations are sedimentary deposits that consist of layers of iron oxides and layers of $SiO_2$. Smelting the iron out of the ore requires heating the ore to very high temperatures in the presence of carbon (usually coal). This process forms molten iron that can be poured out of the smelter and leaves a residue of silicate "slag." This required combination of iron ore and coal made cities such as Pittsburgh an important center for iron and steel production in the late 19th century (Figure 4.4).

The molten iron can be mixed with more carbon and with other metals to form different kinds of steel. Hundreds of kinds of steel are manufactured for different industrial uses. They differ primarily in the amounts of various alloy metals. We discuss eight of the most common alloy metals without attempting to describe the properties of the steels they are used in.

Chromium is mined almost entirely as the mineral chromite ($FeCr_2O_4$). Chromite accumulates in large intrusions of basaltic rocks. The principal deposits are in South Africa, Kazakhstan, and India.

Cobalt forms in veins as the mineral cobaltite (CoAsS). The major mines are in a long belt of mines in Congo (Kinshasha) and Zambia. Significant production also occurs in Russia, Canada, and Cuba.

Manganese is a sedimentary deposit of oxides and hydroxides similar to the minerals of iron ores. It is produced mostly in South Africa and Australia, with minor production in Gabon and the United States.

Molybdenum occurs as the mineral molybdenite ($MoS_2$) and is mostly mined as a byproduct of some copper mines. Major production is in the United States and Chile.

Nickel occurs mostly in veins as the minerals niccolite (NiAs), pentlandite ($Fe,Ni_9S_8$) and pyrrhotite ($Fe_{0.83-1}S$). One of the world's largest deposits is at Sudbury, Ontario, and some geologists propose that the nickel is from a huge meteorite that impacted there 1.8 billion years ago. In addition to Canada, nickel is produced at a Russian mining complex at Norils'k in northern Siberia. Because of smelter fumes and large waste dumps in the city, Norils'k is regarded as one of the most polluted cities in the world.

Titanium occurs as rutile ($TiO_2$). It is mined from large bodies of intrusive rocks, principally in Australia and South Africa.

Tungsten occurs in veins mostly as wolframite ($Fe,MnWO_4$). The major production in the middle 20th century was centered around the state of Nevada, but virtually all of the world's tungsten now comes from China.

Vanadium occurs in veins as a variety of minerals. Almost all of it is produced in South Africa and Russia.

### Nonmetals

The volume of nonmetals mined is immensely greater than the volume of metals. Nonmetals range from large slabs of building stone to tiny gems. They include gypsum used in wallboards and plaster and also salt used as a seasoning.

Most of the nonmetals mined are used to make buildings and roads. Both of them start with crushed rock.

Crushed rock is made by blasting rock out of quarries and running it through milling equipment. One of the first uses of crushed rock was to put it on top of dirt roads and convert them to roads that could be used in wet weather. This process was first developed by Scottish engineer John L. MacAdam in the early 19th century. MacAdam's method consisted of covering dirt roads with about 6 inches of crushed gravel and then compressing the gravel with a roller so that it formed a stable and coherent mass. When the road surface was made even smoother by covering the gravel with a layer of asphalt (tar), the result was "tarmac."

Crushed rock and cement together make concrete. The cement is made by mixing crushed limestone ($CaCO_3$) with sand and heating them until they are "calcined." Calcining develops a type of calcium silicate that reacts with water to "set" to a hard mass unless it is constantly stirred. Consequently, when cement and crushed rock are combined to form concrete and then mixed with water, trucks that carry concrete to a construction site consist of large round barrels that are constantly rotated to prevent setting before the concrete is poured.

### History of Tools

Chimpanzees are humans' closest relatives and have been observed picking up branches or rocks to use as primitive tools. The earliest humans also used wood and rock as long ago as 1 million years.

The first technological advance made by humans was control of fire, which no chimpanzee has ever accomplished. Archaeologists have demonstrated the use of controlled fire by finding remains of fire pits. Most of these pits were constructed by arranging stones in a circle and building fires in the center. Any charred material that remains in these pits can be dated by methods that we describe in Chapter 3.

The age of the oldest fire pit is controversial, with estimates ranging from several hundred thousand years to more than 1 million years. This evidence shows that primitive humans learned to use fire before modern people (*Homo sapiens*) evolved in Africa about 100,000 years ago.

*Homo sapiens* and Neanderthals, who are closely related, continued to rely solely on wood, stone, and fire until the Cro Magnon culture evolved. Their fossils were first found in southern France in an archaeological site about 25,000 years old. Cro Magnon were clearly the ancestors of modern humans, and they began to fashion tools considerably more sophisticated than anyone who lived before them.

### *Types of Rock Used for Tools*

All people who made stone tools had to select rocks carefully. Arrowheads and other tools with sharp edges must be made from very different rocks than are used for stone pots. Sharp edges can be fashioned only on rocks that are hard and brittle. Volcanic rocks that have high $SiO_2$ contents (rhyolites) are ideal, and many sharp tools were made from rocks formed by explosive volcanic eruptions (see Glowing Clouds in Chapter 2). Other rock types that are useful include chert, which is known as flint if it is black. Excellent tools can be made from obsidian, a black volcanic rock that cooled so rapidly when it was erupted that it formed glass. The mineral quartz makes good arrowheads, but it is very difficult to fashion.

Some rocks are too soft to be used to make sharp edges. One example is basalt, a volcanic rock with low $SiO_2$ that becomes even softer when it "weathers" by exposure to air and water. Soft rocks were used largely to make bowls and "bannerstones." Bannerstones are stones with holes in the middle so that they can be slid over narrow shafts of wood. Shafts fitted in this way are called "atlatls." Hunters could throw spears that were balanced on the ends of atlatls, and the bannerstone would give greater momentum to the spears than the hunters could generate just by using their arms.

### Stages of Tool Development

The ability to make and use tools continually evolved. Archaeologists and historians divide this evolution into several stages, and we discuss them here.

*Paleolithic.* The Paleolithic period refers to a time when early humans used only crudely fashioned stone tools. Some of the tools were hand axes and choppers, which were simply rocks with a sharp edge. Many of these rocks may have had naturally sharp edges when they were cracked off of outcrops. Some, however, may have been sharpened by "flaking" through hitting the rocks with other stone tools. Flaking produced tools known as "bifaces" because both sides of the rock taper to a sharp ridge. Paleolithic societies gradually came to an end when modern humans (Cro Magnons) became dominant and developed better stone tools in Neolithic time. Archaeologists generally date the start of the Neolithic at about 10,000 BC in the Middle East, but the transition did not occur at the same time in all parts of the world. Paleolithic societies had developed worldwide, and the farther they were from the Middle East, the later was their transition to Neolithic societies.

*Neolithic.* Major changes in the early Neolithic at about 10,000 BC were not just more sophisticated stone tools, but also the development of sedentary societies that replaced the older hunter-gatherer lifestyle. Planting food crops and herding domestic animals began in the Middle East by about 8000–7000 BC. Planting crops began in Central America by about 5000 BC and spread northward and southward by about 3000 BC (Archaic time in North American chronology).

Raising food by planting crops and herding animals gave people more leisure to devote to other innovations. Wheels were invented in the Middle East before 5000 BC. Wheels not only aided transportation but also the crafting of pottery, and an economy based on "consumer goods" became more common.

*Copper Age.* People in the Middle East learned to smelt copper by about 4000–3000 BC, and the technology spread to Europe and Asia within 2,000 years. Small deposits of copper ores are widespread, and there was never any difficulty in finding enough copper. Rich deposits were naturally more useful, and the island of Cyprus had so much ore that copper mined there was distributed throughout the eastern Mediterranean and Middle East. Cyprus was still so famous in Roman times that our name for copper and its chemical symbol (Cu) comes from "Cuprum," the Roman name for Cyprus.

The inability of Native Americans to smelt any metals other than gold and silver made copper available to them only where they found deposits of "native copper" (raw metal). These deposits are primarily along the southern shore of Lake Superior, and the difficulty of mining them restricted the use of copper mostly to small items of jewelry.

Where copper was abundant, it rapidly replaced stone for many uses. Copper was fashioned into axe heads, spear points, cooking pots, and bowls for eating and drinking. The greater ease of making these implements from copper instead of stone gave people more leisure time. They used this time to develop writing, better vehicles, and a more organized (hierarchical) society. Better transportation encouraged people to move into cities and develop a large number of artisans who were fed by people who remained on the land to produce food.

*Bronze Age.* Native Americans completely missed the Bronze age, but in the Middle East and other parts of western Asia it began at about 3000 BC. Early artisans working with copper realized that the solid copper was easier to shape and the molten copper was easier to pour if it contained small amounts of other metals. Modern analyses of these early artifacts shows that the most common "impurities" were lead and antimony, which probably came from minerals naturally associated with the copper ores.

These metals did not improve the physical properties of copper, but when workers added a few percent of tin they produced a superior metal called bronze. Bronze was easier to shape than copper and was also harder and held a sharper edge. By about 3000–2000 BC, bronze tools had replaced copper tools wherever artisans could obtain tin.

Tin production in the Bronze age came from Southeast Asia and Cornwall, England. The Cornwall mines are mined out now, but they were a major source of tin during Roman times and had begun shipping tin to the Mediterranean before the Romans occupied England.

*Iron Age.* Iron began to be produced in the Middle East by about 2000 BC or slightly later. Although small iron deposits are available almost everywhere, iron could not be smelted until people learned a process far more complex than smelting copper and tin. Metallic iron must be separated from its ore by heating the ore to very high temperatures in the presence of carbon. Then early artisans had to remove more impurities by folding the solid iron repeatedly ("working" it).

When iron became plentiful, it replaced bronze in almost all of its earlier uses. Iron became essential for axes, knives, and utensils. Also, armies that were equipped with iron weapons could easily dominate those that depended on the far less abundant bronze.

As with copper, iron was used in the Middle East before it was used elsewhere. Thus, historical events are described as occurring in the Iron age even in places where iron had never been used. In Scotland, for example, people still depended on stone instead of iron for protection, and they produced "vitrified" forts by building rock walls and then partially fusing them with fires set along the walls.

### Fertilizer

Plants grow only if they are well supplied with nutrients. Traditional farming provided these nutrients through a combination of plant mulch and the dung of grazing animals. The Green Revolution (see Food Supply and Population Growth) was possible only with the aid of artificial fertilizers, which are also routinely sold to home gardeners.

The three major nutrients in fertilizer are nitrogen (N), phosphorous (P), and potassium (K); the nitrogen is generally a mixture of nitrate and ammonia. Fertilizer is commonly marketed with three numbers that show the concentration of the major nutrients. For example, a 10–12–8 fertilizer contains 10 percent total nitrogen, 12 percent phosphate, and 8 percent potassium.

The components of fertilizer come from various sources. Nitrogen compounds are made by converting atmospheric nitrogen to ammonia or nitrate. Phosphate is produced by mining phosphate rocks and then combining them with acid to create "superphosphate," which is more soluble than the rock phosphate. Potassium is mined as various forms of potassium minerals and then converted to potassium chloride or potassium sulfate.

Rock phosphate occurs in many places and is formed by various processes. One of the strangest sources is Nauru, a tiny island in the western Pacific and home to an enormous number of birds. Bird droppings are rich in phosphate, and they are so abundant on Nauru that more than 75 percent of the island has been mined out and converted to an uninhabitable wasteland.

When artificial fertilizer became popular in the late 1800s, the process of making the nitrogen component from atmospheric nitrogen had not yet been discovered. It was, therefore, necessary to search for nitrate minerals. One of the richest sources was in the extremely dry Atacama Desert on the west coast of South America. The area was originally owned by Bolivia, which had a port on the Pacific Ocean. In a war that lasted from 1879 to 1884, however, Chile took over the nitrate deposits and left Bolivia landlocked.

The modern fertilizer industry is surveyed by the International Center for Soil Fertility and Agricultural Development (IFDC). Their summary statistics show that more than 500 million tons of fertilizer was sold worldwide in 2006.

### Precious Metals and Gems

Precious metals and gems have always been used for ornaments and as a form of money. We illustrate the problems that they pose by discussing the search for gold and the production of diamonds.

### *Gold*

A mine is:

"a hole in the ground with a liar at the top;"
"a hole in the ground with a fool at the bottom."

These quotes are attributed, perhaps incorrectly, to Mark Twain, and they describe gold mines more than mines of any other ore.

Throughout human history the beauty of gold has made it sought after for ornaments and jewelry. This value also made gold the equivalent of money; a common means to pay for commercial transactions before societies began to manufacture any forms of money. Most of this money was then valued in terms of how much gold it represented.

The United States was on a "gold standard" through the 1800s and into the 20th century. All money printed at this time was backed by the amount of gold in government storage, principally at Fort Knox, Kentucky. This standard meant that anyone who had a dollar bill could, in theory, go to the U.S. Treasury and receive one dollar's worth of gold. In order to maintain this exchange policy, the U.S. government had to place strict limits on the amount of paper money and coins in circulation. The policy also meant that much of the gold taken out of mines was immediately put back underground in storage vaults.

**Figure 4.5** Gravel bar near Sutter's Mill.

Limiting the amount of money available to the public kept inflation low but placed great hardship on people who were poor and in debt. There was so little money in the country that poor people couldn't get enough to pay their debts or even buy the necessities of life.

The hardship caused by the limited amount of money available under the gold standard became a political rallying cry. At the 1896 Democratic National Convention, William Jennings Bryan advocated basing the supply of money on both gold and silver and said, "You shall not crucify mankind upon a cross of gold!"

The value of gold has always meant that people hurried to get their hands on it when a new source was discovered. One of the most famous "gold rushes" in history was spurred on January 24, 1848, when James W. Marshall discovered gold in gravel at a lumber mill owned by John A. Sutter (Figure 4.5). The mill was on the American River east of San Francisco, and gold was discovered by many other people at various places along the river at the same time. News of the discovery reached people in the eastern part of America and the rest of the world by the end of 1848.

Prospectors looking for gold were called "Forty Niners" because they began to arrive in San Francisco in 1849. By 1855 several hundred thousand had come, not only from the United States but also from other parts of the world. Some of them from the eastern United States took several months to come by wagon, which required crossing the desert of Nevada and the Sierra Nevada Mountains of California. Some took 3 months on clipper ships around Cape Horn. Some opted for the fastest route, taking about 1 month to sail from New York to Panama, walk across the isthmus, and take another ship north to San Francisco (see Panama Canal in Chapter 1).

When the prospectors arrived at the American and other rivers flowing west from the Sierra Nevada, they looked for gold in the sediments deposited by the rivers. Gold is just one of several minerals that are heavy (dense) enough to fall to the bottom of streams and accumulate as "placer deposits."

The simplest way to extract placer gold from the other sediment is by "panning." A pan looks very much like a Chinese wok used for cooking. A miner put a little sediment and water in the pan and then rotated it rapidly. This motion swirled the lighter sand and gravel over the side of the pan and left the heavier flakes of gold ("color") on the bottom.

It was immediately obvious that gold in the rivers flowing out of the Sierra Nevada came from eroded gold-bearing ores in the mountains. These vein deposits were referred to as the "mother lode" and were actively sought by prospectors. Ultimately several lode gold mines were discovered in the Sierra foothills slightly east of a route that is now appropriately followed by California State Highway 49.

The Forty Niners were neither the first nor the last people who looked for gold. The earliest may have been a real (or perhaps mythical) Greek named Jason. Jason is said to have taken a ship manned by a band of adventurers, called Argonauts, and set out to find gold at some time near 2000 BC (give or take a few centuries). Jason may have looked for it on the southern shore of the Black Sea, where placer gold is known to have been produced since the very early days of human history.

The method of extracting this gold consisted of placing skins of sheep in the rivers to trap the very fine gold dust being washed along by the water. When enough gold had accumulated on the skins, people would take them out of the river and comb the gold dust off. A sheepskin with enough gold to shine was called a "golden fleece."

Fleecing in a different sense is still a time-honored tradition in gold prospecting. False claims of gold deposits and defrauded investors are common throughout history. One of the largest frauds in recent history involved Bre X, a small Canadian company with no experience in international scheming.

In 1993 Bre X hired Michael de Guzman, a Filipino geologist, to investigate a reported gold deposit they had bought in central Borneo. de Guzman reported assays that showed a reserve that ultimately climbed to 200 million ounces in 1997. At this point, other mining companies considered investing in the mine, and the Indonesian dictator, H.M. Suharto, wanted his share.

Naturally, other mining companies wanted confirmation of the assays before investing, and they sent their own geologists to confirm the original assays. In 1997, 1 week before the new assays were announced, de Guzman fell out of his helicopter when it was about 200 meters above the ground, either because he wanted to commit suicide or because he was pushed.

When the new assays were announced, it became clear that there was no gold at the Bre X property, and the original assays were frauds. The only gold in the samples had been put in by adding shavings of gold from other sources, including jewelry, to the barren rock (geologists refer to this process as "salting" a claim). This revelation caused Bre X stock to drop immediately from its high of $280 per share to almost zero, leaving investors with a total loss of more than $4 billion. The ultimate bankruptcy of Bre X also drove its two major founders out of Canada. One died in the Bahamas in 1998, and the other fled to the Cayman Islands, which does not have an extradition treaty with Canada, in order to avoid criminal prosecution.

Despite all of the shenanigans in the gold business, the industry has continued to grow. In the early part of the 21st century, worldwide consumption and production range from 2,500 tons to 3,000 tons. (At a current price of about $800 per ounce, 1 ton of gold is worth approximately $25 million.) The ranking of major producers is South Africa (1st), Australia (2nd) the United States (3rd, mostly from Nevada), then China, Russia, Peru, and Canada.

The ranking of gold-consuming countries is very different from that of producers. The largest consumer is India, where gold jewelry is much prized as a way for brides to bring dowries to their future husbands. Following India, major consuming countries are the United States, European Union, China, and Saudi Arabia.

The amount of gold produced and consumed is dwarfed by the 20–30 million tons held by the world's central government banks, including the U.S. Federal Reserve. Much of the price of gold is controlled by these banks, which can lower the price by selling some of their reserves or raise the price by buying more gold.

The price of gold does not reflect many of the human costs associated with the industry. Any kind of mining is dangerous, and some gold mines are 3 kilometers deep. Miners are mostly poor people who need jobs even though the work is dangerous, and they commonly live under degrading conditions in company-owned "camps." The refrain, "I owe my soul to the company store" is from a song about miners who have to live in a company-owned town and buy their food at the company grocery.

Another serious effect of the gold industry is that many gold mines are operated in underdeveloped (poor) countries by companies based in developed (rich) countries. The mine operators commonly disregard the rights of native people and let their mining practices destroy the environment that these people require in order to survive.

### Diamonds

Carbon in the earth can crystallize as either graphite or diamond. Crystallization at depths of 100 kilometers or more in the mantle generates diamond, the hardest mineral formed in nature. Diamonds reach the earth's surface in finger-like "pipes" of kimberlite, a rock with the general composition of basalt but containing an unusual assortment of minerals that rarely occur in other rocks. Kimberlite pipes apparently travel at very high speeds from the mantle to the earth's surface. Kimberlite weathers rapidly at the earth's surface, and the diamonds in it can either be mined directly or from placers downstream from the pipes.

Kimberlite is named for Kimberley, South Africa, where the first diamond pipe was found. South Africa and neighboring countries still supply more than half of the world's diamonds, with smaller amounts from Canada, India, Brazil, and Australia.

Legal trade in diamonds is very tightly regulated, but illegal trade has flourished since the latter part of the 20th century. Money from sale of illegal diamonds has been used to support wars and revolutions in several countries, principally Angola, Ivory Coast (Cote d'Ivoire), Liberia, and the Republic of the Congo (former Zaire).

The United Nations defines these "conflict," or "blood" diamonds as "diamonds that originate from areas controlled by forces or factions opposed to legitimate and internationally recognized governments, and are used to fund military action in opposition to those governments, or in contravention of the decisions of the Security Council." The United Nations and legitimate diamond sellers are working very hard to stop the illegal trade in diamonds.

## FOOD

This discussion is divided into three parts: (1) nutritional require-ments; (2) sources of food and their changes through time; and (3) the Green Revolution and the problem of population growth.

### Nutritional Requirements

The amount of food required by individual people varies according to body weight, gender, and level of physical activity. About 2,000 calories per day, however, is a minimum for adults who lead comparatively inac-tive lives (note that food calories are actually kilocalories). The amount ranges upward for heavy people in strenuous activity, and professional football players in the United States regularly consume about 5,000 calo-ries per day. People on diets can lose about one pound (0.5 kg) per week by reducing consumption by 100 calories per day.

Consumption varies among countries and among different groups of people within countries. People in industrialized countries generally consume more than 3,000 calories per day, but those who live in poor countries must survive on not much more than the minimum 2,000 calories per day. Even in the United States, with an average consumption of about 3,500 calories per day, several million children in poor families do not receive an adequate diet.

The number of calories in different foods is variable, with more in meat and less in fruit and leafy vegetables. The average, however, is about 3 cal/g, which is equal to 1,600 calories per pound. Thus, a diet that provides 3,500 calories per day requires a food intake of about 1,000 grams (2 lbs) of food per day.

People have always used food from an enormous variety of sources. Before the development of agriculture (see Neolithic discussed earlier), everyone lived in "hunter-gatherer" societies, depending on animals and fish for meat and on berries, nuts, and wild grain for vegetables. These diets were nutritious despite the difficulties of procuring the food. In addition to calories, they provided essential nutrients that have been absent from some of the more recent diets.

People in Neolithic societies commonly depended on one crop for all of their nourishment—maize in the Americas and wheat or oats in Eurasia. This dependence and the absence of meat for poor people caused most Neolithic people to have serious deficiencies in nutrients such as protein and vitamins $B_{12}$ and vitamin C.

Vitamin $B_{12}$ regulates neurological processes and is essential for the production of red blood cells. It is present only in meat, and people on vegetarian, particularly vegan, diets generally take vitamin pills in order to remain healthy. Vitamin C prevents scurvy. It is present in most vegetables and is synthesized by all animals except humans. Thus, people must consume either vegetables or fresh meat in order to avoid scurvy, which was a scourge of mariners or polar explorers who tried to live on canned or salted meat for long periods of time.

The types of food consumed by people have changed enormously from early agricultural societies to the present. Modern diets are a mixture of vegetables and meat. Meat, particularly beef, is more expensive than grain because people prefer to eat grain-fed ("fattened") cattle instead of the tougher meat from grass-fed, "free-range," cattle. The expense arises not only from industrial processing but also because 8 pounds of grain are needed to produce one pound of beef. Because of its expense, people in rich societies generally eat more meat than those in poor societies.

Two factors are increasing worldwide food prices in the early 21st century. The first is increasing consumption of meat by rich societies, which reduces the supply of grain on world markets. The second factor that also drives up the cost of grain by reducing its supply as a food is increased use of grain for biofuels (see Renewable Resources) Because they eat less meat and use less fuel, the world's poor people bear almost the entire cost of these increasing prices.

**Sources of Food**

Modern transportation has now made food from around the world available to almost everyone. Types of food were far more restricted, however, before Europeans began arriving in the Americas in the 1500s. This contact ended an isolation that had existed for more than 150 million years as the Americas drifted away from Europe and Africa to form the Atlantic Ocean (see Peopling of the Americas in Chapter 1). During this separation, the Americas evolved a fauna and flora entirely different from that of the Old World. The few exceptions were animals, such as mammoths, which migrated around the polar regions during the ice ages.

Europeans brought *to* the Americas:

*Horses.* Some types of horses evolved in North America. They became extinct, however, shortly after the last ice age, and all modern horses are descended from ones brought from Europe. Because horses were expensive, however, the colonists soon brought donkeys and produced mules as the sterile offspring of crossbreeding horses and donkeys. Some of the horses brought by early Spanish explorers either escaped or were stolen by Native Americans, and their descendents bred so rapidly that Native Americans soon had an abundant supply of horses for riding and carrying goods.

*Domestic farm animals,* such as cows, pigs, goats, sheep, and chicken.

*Wheat,* which arrived in several waves. Most of the early types of wheat were "soft" varieties that matured during the summer and were harvested in the fall. They are relatively low in protein and are best used for bread. Later, people brought "hard" varieties (durum wheat) that mature over the winter and are harvested in the spring. They are richer in protein and best used for making pasta. The final wheat to arrive was a hard Ukrainian strain known as Turkey Red, which arrived with Russian Mennonites in 1873 and has proved so reliable that it now accounts for almost all American wheat.

*Beans and other legumes.* Europeans brought legumes that had evolved largely in the Mediterranean area to join the beans of the Americas (genus *Phaseolus*). The imports included broad beans (genus *Vicia*) plus such related varieties as chickpeas (garbanzos), and green peas. Because the broad beans are not as tasty as Native American varieties, they have never been as extensively cultivated, but peas have become a staple of American diets. The third of the major bean genera is *Glycine* (soybean), which was imported from its home in East Asia and has now become the major bean produced in North America.

*Fruit.* Apples, pears, peaches, and several varieties of melons arrived with European settlers, although not all before the Revolution. Oranges, lemons, and other citrus fruit developed in India and Southeast Asia and were not brought to the Americas until growers began extensive cultivation of groves in Florida in the 1800s.

*Rice.* The rice that was such an essential crop for colonists in the Atlantic coastal plain before the Revolution arrived in the Americas by accident. Originally from India, it was taken to Madagascar by Arab traders and cultivated there. Then, in the early 1600s as rice was being shipped around the world, a vessel carrying it was wrecked on the coast of South Carolina. After the crew was rescued by local people, the captain gave them two bags of rice that proved to be the seed material for the development of the entire rice industry in the southeastern United States.

*Tobacco.* The type of tobacco grown from colonial time to the present arrived by a circuitous route. Native Americans had long cultivated tobacco, but Europeans who imported it found it greatly inferior to varieties that grew in the Caribbean. Consequently, colonial planters brought in the

Caribbean tobacco, and the native variety has never been used since then.

*Cotton.* Some types of cotton evolved in North America, but the types we grow now were taken back to Africa and then returned to North America. Cotton was grown extensively by colonists in the coastal plains, but the invention of the cotton gin in 1793 by Eli Whitney allowed cotton to be grown in the southern piedmont region.

The only animal that Europeans took *away* from the Americas was the turkey, but the list of plant foods distributed to the rest of the world is impressive.

All varieties of *Phaseolus* beans, a genus that includes beans such as lima, green (string), navy, pinto, and kidney.

Vegetables such as squash, zucchini, and tomatoes.

Maize, which is referred to as corn in America.

Potatoes. They are natives of South America, were taken to Europe, and then brought back to North America.

Chocolate and avocados from Central America.

Peanuts (groundnuts) were taken from South America to Africa and then returned to North America by slaves.

Now the marketplace for food is global. Virtually every country in the world uses food native to the New World and food native to the Old World. On a weight basis, the largest crop produced in the world is sugar (both from cane and beets). Sugar is followed in sequence by wheat, maize (corn), rice, and potatoes.

## Food Supply and Population Growth

Thomas Robert Malthus couldn't have known how wrong he was in 1798. Or how much damage he would do to the lives of the billions of people who have lived since 1798. Malthus' *Essay on the Principles of Population* seemed mathematically reasonable when he published it in 1798. He had noticed that the human population tended to increase "exponentially" because the number of children born each year is proportional to the number of people of childbearing age. (Actually births are proportional to the number of women of childbearing age, and the number of men is irrelevant.) So as more girls grew to childbearing age there would be more children born, and then these new women would have even more children to yield an increase in population that would become faster and faster.

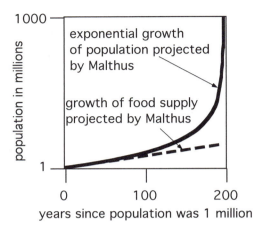

Based on this concept, a graph of population through time looks like the one in Figure 4.6. The rate of increase can be described in terms of "doubling time," which is the number of years in which the population doubles from its size at the start of the time period. For example, a country that has a population of 10 million now and a doubling time of 20 years will have a population of 20 million in 20 years. Then it will have a population of 40 million in another 20 years (40 years from now) and a population of 80 million 60 years from now. This concept of doubling time is exactly the reverse of the "half life" of radioactive elements (see Chapter 3).

**Figure 4.6** Malthus' prediction of growth in population compared to growth in food supply.

Malthus proposed that this exponential increase in the population would outstrip the increase in food supply, which would grow "linearly" (Figure 4.6). This "Malthusian doctrine" implied that the increase in population would lead to starvation, and people in this growing population would presumably fight each other, either as individuals or as societies. The idea that people would have to fight in order to survive has been used by leaders in many countries to justify the concept of "survival of the fittest" ("natural selection"), a concept developed by Charles Darwin in 1859 to explain the evolution of new types of animals and plants and the extinction of unsuccessful forms (see Chapter 3).

Estimates of world population are unreliable until the last few decades, but the population is reasonably well-known for the last 1,000 years (Figure 4.7). The graph shows an accelerating increase in population throughout this period, with a decrease in doubling time from several hundred years in 1000 AD to fewer than 100 years now.

The continued increase in population is seldom affected by two events—war and epidemic—that are traditionally thought to control population. For example, the population of Europe was interrupted only briefly by two episodes of plague ("Black Death") but not by any of the numerous wars that convulsed the continent. Even the deaths of about 30 million combatants and civilians during the World War II only slowed, but did not reverse, the trend toward an increasing population, and Europe contained more people at the end of the war than at its start.

During the past 50 years, world population has increased by more than 1 billion people as lifespan increased and fewer children died. But how could this happen if Malthus was right and there isn't enough food for people to keep themselves alive? The answer is increased food supply caused by the "green revolution."

The most recognizable person among the developers of the green revolution is Norman Borlaug, who received a Nobel Prize in 1970. Many

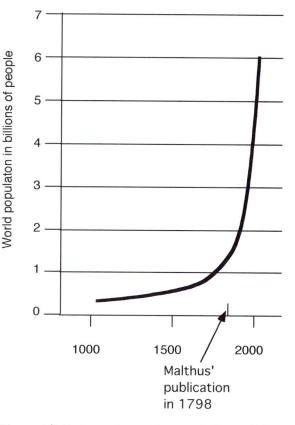

**Figure 4.7** Estimated world population of last 1,000 years.

other people doing agricultural research, however, also contributed to the efforts to increase crop yields and allow farmers to grow more food. The result is that world food production has increased tenfold in the past 50 years, a rate much faster than the rate of population growth.

This increase in production was accomplished by several changes in agricultural methods. The use of both natural and artificial fertilizers, coupled with better design of water systems, caused plants such as wheat to grow more rapidly and develop many more grains of edible wheat on a single stalk. Unfortunately, this greater number of grains caused wheat stalks to become top-heavy and fall over. Research workers solved this problem by developing "semidwarf" plants with shorter stalks that could support the larger number of grains.

The environmental effects of the green revolution are now becoming more apparent. Extra fertilizer causes pollution of surface and ground water. Growing wheat in areas that were once covered by grass causes increased soil erosion. Some plant varieties that were genetically modified to be resistant to pests now require more pesticides.

Perhaps the greatest problem of the green revolution has been the unequal availability of new seeds and fertilizer to different farmers. Both seeds and fertilizer are expensive, and not all farmers were able to buy them when they first became available. This difficulty tended to separate farmers into "haves" and "have-nots," with the result that economic differences between people became larger.

But overall, we now have plenty of food in the world to feed everybody. So does this mean that nobody is hungry? No. Food is a weapon. We could list numerous examples, but three will suffice.

When crops failed in North Korea in the 1990s, the military leaders of the country preferred to watch some of their people die rather than let outsiders bring in food aid and get a good look at the leaders. In the early 2000s, leaders of Sudan sent paramilitary groups into western Sudan (Darfur) in order to starve out people who did not agree with the leaders. And in the United States, some people are hungry because they do not have enough money to buy the food that is readily available on the shelves of markets.

Despite the abundance of food, many people continue to believe in the Malthusian doctrine of inevitable hunger as a result of population growth. This belief supports the concept of "survival of the fittest," which can be interpreted to mean that people who are hungry—perhaps starving—do not deserve to eat enough food to survive. That these people deserve to die off (become "extinct") as they are replaced by people who are more entitled to survive.

It is clear that Thomas Malthus is not responsible for hunger in the modern world. Unfortunately it is also clear that his doctrine is used by people who are responsible for modern hunger.

## WATER

People need to drink about 2 liters (2 quarts) of fresh water each day or obtain moisture by eating fresh fruit, such as melons.

Methods for obtaining water have evolved remarkably through time. Early people had no problem finding their water because they lived near rivers and lakes and could scoop water out of them with their hands. People began to dig wells at about the time that agriculture started (see Neolithic discussed above). They also dug ditches to channel water from rivers to farmland.

Romans built aqueducts throughout their empire. They were a combination of tunnels through mountains and arched structures across valleys, and some of them were more than 100 kilometers long. Some of these aqueducts still stand and are major tourist attractions. At their peak, aqueducts delivered more water per capita to the city of Rome than residents of many modern cities receive now.

Modern distribution of water is highly variable from place to place. Some people in the deserts of Africa and the Middle East live around oases, where water is at the surface. Some people, however, haul water out of wells and distribute it to their homes and livestock in jugs and pots (most of the carrying is done by women, who balance the pots on their heads). Small islands commonly do not receive adequate rainfall and either import water (e.g., the Virgin Islands in the Caribbean) or "desalinate" seawater (e.g., Malta). Australia has made major plans to desalinate seawater to overcome the increasing aridity of the entire continent.

Differences in supply and distribution of water lead to extraordinary differences in consumption. Per capita consumption in America is about 40 cubic kilometers per day ($10^{27}$ liters). About two-thirds of that consumption is used for agriculture, one-third for industry, and only a tiny fraction for domestic use and human consumption. By contrast, people living in the Republic of the Congo (former Zaire) use about 0.1 cubic kilometers per day for all purposes, including agriculture.

## WIRELESS COMMUNICATION AND THE INTERNET

The history of communication is a story of continual increase in speed—from slow and uncertain in antiquity to instantaneous and verified today. We start by describing the difficulty of exchanging letters along the Silk Road between China and Rome.

### Silk Road

Silk is produced by boiling and unwinding threads from the cocoons of silk worms, and when China was the only place in the world that had silk worms, anyone else who wanted silk had to buy it from the Chinese.

That monopoly was in effect in the first century BC, when silk first reached Rome. The Romans knew it came from a country ruled by the Qin (Chin) dynasty, and they named the country China. They, and everyone else, also knew that it was important to set up silk industries outside of China in order to reduce prices.

By about 500 AD the Chinese had lost their monopoly on silk. Then other items of commerce began to be more important than silk. The

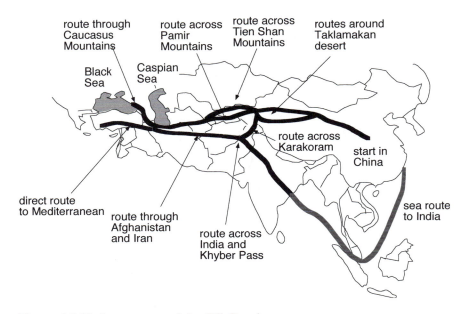

**Figure 4.8** Various routes of the Silk Road.

name remains, however, and even today the route is still called the Silk Road.

The Silk Road isn't one road but many different routes (Figure 4.8). Chinese ships sometimes took silk to eastern India and traders took it over land from there. That route took them across the Khyber Pass, now on the border between Pakistan and Afghanistan.

The eastern end of the Silk Road on land starts west of Xian, China. Xian is an agricultural area, but not as lush as the region around Beijing. West of Xian the land becomes more arid near the western end of the Great Wall of China. Here the wall is made of hardened mud and is little more than a few feet high. During the height of Chinese domination, before invasion by the Mongols, the wall separated the agricultural economy of China from the nomadic economy of Mongolia.

Genghis Khan ended that division in the 13th century. Along with his Golden Horde of horse-mounted warriors, Genghis destroyed the Chinese dynasty and installed his son Kublai on the throne of China. Then, slaughtering without limit, Genghis and his descendents led the Horde all the way to Poland and Hungary before it was so weakened that it withdrew eastward in 1241.

Farther west of Xian is an area of interior basins, some of them containing salty lakes and marshes, and farther west is the eastern edge of the Taklamakan desert (Xinjian province of China). This desert is an area of impenetrable sands, and travelers had to go either north or south.

Tibet and areas to the north are arid because the Himalayas prevent monsoon winds from the Indian Ocean from reaching central Asia. This blockage is so effective that the aridity extends across the Tienshan Range on the northern edge of Xinjiang, through the arid plains of Kazakhstan, and stretches all the way to the forests of Siberia, known as the "taiga." Although the taiga is a forest, it is wet only because Siberia is so cold that the meager precipitation doesn't evaporate. The general aridity of the forest is emphasized by the fact that there was not enough moisture in Siberia to form continental glaciers during recent glacial periods (see Chapter 1).

The routes both south and north of the Taklamakan desert converge at the town of Kashgar, at the western end of the desert. Several routes lead farther west. One is across the Karakoram Range into northern Pakistan. Now the Karakoram highway links China and Pakistan via the highest paved road in the world. It crosses a pass with an elevation of almost 5,000 meters. The route from Pakistan is through the Khyber Pass into Afghanistan, which was created in the late 19th century as a buffer zone between the British and Russian empires, and then through Iran and Iraq into Turkey.

Other routes followed by the Silk Road lead directly west over very high passes in the Pamir Range or slightly north across the Tien Shan Range (also known as the "Celestial Mountains"). Across the Pamirs is Tajikistan, and crossing the Tien Shan Range leads to Kyrgyzstan.

When westbound travelers crossed the Amu Darya (Oxus) River, they entered the parched land of Turkmenistan. Turkmenistan extends westward to the Caspian Sea, which occupies an enclosed basin fed by the Volga River. Since water in the Caspian basin has no outlet, the sea is becoming saltier because of evaporation. This salty water is well suited to the growth of sturgeons, whose roe was the original, and still prized, caviar.

Travelers could go around the Caspian Sea either to the north or south. To the north they would cross the Volga, and enter the fertile lands of southern Russia and Ukraine. Traders generally went south, however, and came to Baku, Azerbaijan, on the western shore of the Caspian Sea. From there they crossed the Caucasus Mountains through the present country of Georgia to the Black Sea (see Figure 4.3 on page 127).

### Development of Modern Communication

Communication was easier within both the Roman and Chinese empires than between them. They both had road systems where relays of horse riders could travel about 150 kilometers a day. Ships with sails in the Mediterranean could carry letters around the Roman Empire at about the same speed, but piracy and storms prevented some of them from reaching their destinations.

Speeds of about 150 kilometers per day both on land and sea prevailed into the 1700s. Neither ships nor horses could move faster until new technologies were developed. They came rapidly in the late 1700s and throughout the 1800s.

The first technical achievement did not change speed, but it made accurate navigation at sea possible. Navigators had always been able to determine their latitude from the elevation of the sun above the horizon, but they could only guess at longitude. In 1761 John Harrison, an English clockmaker, designed watches that would keep accurate time even on ships tossed in storms. By comparing the times on the watches with local time determined from sunrises and sunsets, navigators were finally able to measure their longitude.

Travel by both sea and land became faster in the late 1700s and early 1800s. Ocean travel improved in the 1800s, particularly with the construction of "clipper" ships that could sail at up to 20 kilometers per hour in favorable wind. On land carriages were constructed better and roads were improved so that, by the middle 1800s, some horse-drawn carriages could travel at speeds of 10–12 kilometers per hour. Using relays of horses and drivers, therefore, carriages could travel more than 200 kilometers per day. When the "Pony Express" was started in 1860, relays of horses and riders could send mail to and from the west coast of the United States at about 15 kilometers per hour.

All of these developments were overtaken by completely new technology in the 1800s. Usable steam engines were built by James Watt and others in the late 1700s and installed in railroad locomotives in the early 1800s. These first trains could achieve speeds up to 50 kilometers per hour, and they spurred the construction of railroad tracks in most of the world. Horse-drawn carriages were soon replaced by trains where tracks had been laid.

Development of electricity generators had an even greater effect on communication. Michael Faraday and others built the first reliable generators in the 1830s. They led Samuel F.B. Morse to develop telegraph systems based on the "Morse Code" in the 1840s. The first telegraph systems were local, and it wasn't until 1861 that a telegraph wire connected the east and west coasts of the United States. This connection immediately put the Pony Express out of business after only 1 year of operation. Telegraph connection between Europe and North America was achieved in 1866, when a cable across the Atlantic Ocean connected Ireland and Newfoundland (see Chapter 3).

The next major breakthrough in communications was development of the telephone in 1876. It is generally credited to Alexander Graham Bell, although several other people may have produced similar phones earlier. All early telephone messages had to travel by wire, and calls between Europe and North America did not become possible until a phone cable connected Newfoundland and Ireland in 1955.

Wireless communication (now called radio) was developed locally by several people in the 1890s. The first transatlantic signal was received from Europe in 1902 by Guglielmo Marconi at a station in Newfoundland (see Chapter 3). All initial radio communication was by Morse code until voice messages became possible in 1915.

Except for radios, all communication was through wires (cables) until late in the 1900s. At that time, cell phones and similar instruments expanded opportunities for instant communication over most of the world. This expansion occurred at the same time as the development of e-mail via computers and, in 1991, the Internet.

The Internet was invented by British mathematician Timothy Berners-Lee. In an effort to move large amounts of information instantaneously, Berners-Lee developed servers and coined terms such as: URL, universal resource locator; http, hypertext transmission protocol; www, World Wide Web; and html, hypertext markup language.

We now live in an age when information and messages can be sent around the world with exceptional ease. This ability not only stimulates intellectual and economic activity but also advances political and social freedom.

Social and political freedom results from the inability of dominating people to keep information from people they want to control. Throughout history, despots have ruled by preventing their subjects from learning what is happening in free countries. An outstanding example is construction of a wall completely surrounding West Berlin in 1961. It wasn't there to keep West Germans out of East Germany. The purpose was to keep East Germans from escaping to freedom.

# Units and Abbreviations

Time
   Ma = mega-annum; millions of years before the present
   Ga = giga-annum; billions of years before the present
Length
   1 centimeter (cm) = 0.3937 inch (in)
   1 kilometer (km) = 0.6214 mile (mi)
Area
   1 square kilometer (sq km) = 0.3861 square mile (sq mi)
Volume
   1 liter (l) = 1.057 quart (qt)
   1 barrel (bbl) of oil = 42 gallons (gal)
Weight/mass
   1 kilogram (kg) = 1,000 grams (g) = 2.205 pounds (lbs) = 35.27
      ounces (oz)
   1 kg of water = 1 liter
Energy
   1 calorie (cal) = 4.187 joule (j); 1 cal is the amount of heat needed
      to raise the temperature of 1 gram of water by 1°C
   1 British thermal unit (Btu) = 252 cal
   1 quad is an abbreviation for 1 quadrillion Btu
   1 kilowatt-hour = 859.8 cal
Power (energy per unit time)
   1 watt = 1 joule/sec = 0.2388 cal/sec
Food
   1 food calorie = 1,000 cal as defined earlier
   1 g of carbohydrate generates about 3 food calories when consumed

# GLOSSARY

**Abyssal plain.** The plain underlying oceans at a depth of about 8 kilometers.

**Active margin.** Continental margin where subduction occurs.

**Anticline.** Upward bend of sedimentary layers.

**Asthenosphere.** Mobile part of mantle between core and lithosphere.

**Atoll.** Island made by circular reef.

**Basalt.** Low-$SiO_2$ rock formed by cooling lava.

**Basalt plateau.** Thick suite of basalts.

**Biofuel.** Fuel made from plants that are living or recently dead (can also be made from animal fat).

**Caldera.** Bowl-shaped area of volcanism; commonly a few tens of kilometers in diameter.

**Carbohydrate.** Organic compound that consists of carbon, hydrogen, and oxygen.

**CFC (chlorofluorocarbon).** Compound consisting of carbon, fluorine, and chlorine.

**Chlorophyll.** Catalytic compound in plants that uses sunlight to convert $CO_2$ plus $H_2O$ into carbohydrate,

**Continental shelf.** Area of shallow water that extends seaward from shoreline.

**Continental slope.** Underwater area that connects continental shelf to abyssal plain.

**Core.** Central part of earth (approximately 1/8 of volume)

**Coriolis effect.** Deflection of wind directions by rotation of the earth.

**Croll-Milankovitch.** See Milankovitch cycles.

**Crust.** Outer part of the lithosphere.

**Deglaciation.** Melting of glaciers at the end of an Ice Age.

**Depleted uranium.** Uranium after removal of most of the $^{235}$U.

**Discontinuity (seismic).** Surface between two zones with very different velocities of seismic waves.

**Distributary.** Stream that flows away from a main stream.

**Doubling time.** The time needed to double the size of an existing population.

**Drainage basin.** Area in which all tributaries flow toward one main stream.

**Drainage divide.** Line between two drainage basins.

**Drumlin.** Mound of debris smoothed by overflowing glacier.

**Earthquake intensity.** Amount of earth shaking at a point.

**Earthquake magnitude.** Amount of energy released by an earthquake.

**Eccentricity.** Shape of earth's orbit around the sun.

**El Nino.** Weather caused by variation in surface temperatures in the tropical Pacific Ocean.

**Enriched uranium.** Uranium with more $^{235}$U than natural uranium.

**Epicenter (earthquake).** Point on earth's surface directly above earthquake focus.

**Erratic (glacial).** Large blocks deposited by melting glaciers.

**Esker.** Debris deposited by a subglacial stream.

**Floodplain.** Flat area covered by water during floods along a river.

**Focus (earthquake).** Center of energy release during an earthquake.

**Glaciation.** Period of glacial advance.

**Glowing clouds (volcanic).** Ash and steam that flow out of a volcano.

**Granite.** Rock formed below the earth's surface by crystallization of $SiO_2$-rich magma.

**Greenhouse gas.** Gas in the atmosphere that absorbs infrared radiation from the earth's surface.

**Green revolution.** Development of crops that produced more food than unmodified crops.

**Gyre.** Circular oceanic current.

**Half life.** Period of time needed for half of an existing radioactive isotope to decay.

**Hanging valley.** Tributary that enters high on the wall of a deglaciated valley and commonly forms a waterfall.

**Headward erosion.** Process by which streams become longer by eroding upstream.

**Hominin.** Living humans and their ancestors after split from chimpanzees at about 6 Ma.

**Hot Spot.** Area of continued volcanism (commonly tens of kilometers in diameter) that is stationary with respect to Earth's mantle.

**Hydraulic head.** Pressure caused by water at high elevation.

**Hydrocarbon.** Compound that consists of hydrogen and carbon.

**Ice Age.** See Glaciation.

**Induced electrical current.** Electrical current caused by rotation of a magnet near a wire.

**Infrared radiation.** Electromagnetic radiation with wavelengths from $1 \times 10^{-3}$ to $7 \times 10^{-7}$ meters

**Interdistributary area.** Low area between river distributaries.

**Interior drainage basin.** Area where streams flow into a basin and soak into the ground without reaching the ocean.

**Island arc.** Chain of islands with an arc shape.

**Isostasy.** Concept that the earth's surface is floating on the inner mantle.

**Lava.** Magma that flows on the earth's surface after eruption from a volcano.

**Lithosphere.** Rigid part of the earth that consists of the crust and outer part of the mantle.

**Magma.** Molten silicate in the earth.

**Magnetic polarity.** Orientation of the earth's magnetic field. It reverses direction at irregular time intervals.

**Magnetic stripes.** Zones of alternating magnetic intensity on the sea floor caused by magnetic field reversals.

**Mantle.** Region of the earth between the core and crust.

**Meanders.** Broad loops in streams.

**Megafauna.** Large animals.

**Mercalli scale.** See Earthquake intensity.

**Mid-ocean ridge.** See Spreading center.

**Milankovitch cycles.** Orbital variations that include ellipticity, obliquity, and precession.

**MOHO (Mohorovicic discontinuity).** Surface between crust and mantle.

**Moraine.** Deposit at the edge or toe of a melting glacier.

**Natural levee.** Deposit along riverbanks caused by flooding.

**Normal fault.** Fault where blocks on opposite sides pull away from each other.

**Obliquity.** Rocking of the earth's axis with respect to the plane of its revolution around the sun.

**Orogenic belt.** Present or former mountain belt; old belts may be so eroded that they no longer have high elevations.

**Orogeny.** Compression that forms an orogenic belt.

**Ozone hole.** Region of the stratosphere that contains less $O_3$ than normal.

**Passive margin.** Continental margin where no earthquakes or volcanism occur.

**Photosynthesis.** Combination of $CO_2$ and $H_2O$ to make carbohydrate, catalyzed by chlorophyll.

**Placer.** Accumulation of heavy minerals in stream or beach sediments.

**Plate tectonics.** Concept that the earth's surface consists of stable plates surrounded by active margins.

**Plume.** Area of rising mantle. Generally has a diameter of a few tens of kilometers.

**Point bar.** Sediment deposited on inside of meander bend in streams.

**Precession.** Cyclic change in the orientation of the earth's axis of rotation.

**Rifting.** Separation of blocks that pull away from each other.

**Rift valley.** Valley formed by rifting.

**River terrace.** Flat area formed by river before it eroded its valley further; terraces are higher on valley walls than floodplains.

**Saffir-Thompson scale (hurricanes).** Scale that measures hurricane intensity by wind speed.

**Seismic wave.** Method of transmission of earthquake by shaking the earth.

**Shield (continental).** Area where old rock is exposed at the earth's surface.

**Solar cell.** Plate that contains chemicals that convert sunlight to electricity.

**Spreading center.** Ridge in the ocean where plates move away from each other.

**Strike-slip fault.** Fault where blocks move past each other horizontally.

**Subduction.** Movement of lithosphere deep into the mantle.

**Sunspot cycle.** Variation in activity of sunspots, commonly associated with variation in solar intensity.

**Supercontinent.** Assembly of nearly all of the earth's continental crust into one landmass.

**Superphosphate.** Fertilizer made by reacting acids with phosphate rock.

**Syncline.** Downward bend of sedimentary layers.

**Thrust fault.** Fault where one block is shoved over the other block

**Trade winds (easterlies).** Winds blowing to the west in the tropics.

**Transform fault.** Very large strike-slip fault that separates crustal blocks with different geologic histories.

**Transuranic element.** Element heavier than uranium.

**Tree ring.** Ring of wood representing growth added by a tree in 1 year.

**Trench (oceanic).** Very deep area of ocean above a subduction zone.

**Tributary.** A stream that flows into another stream.

**Triple junction.** Place where three rifts come together.

**Tsunami.** Enormous wave in ocean basin.

**Ultraviolet (UV) radiation.** Electromagnetic radiation with wavelengths from $4 \times 10^{-7}$ to $1 \times 10^{-8}$ meters

**Unconformity.** Surface between deformed rocks and overlying undeformed rocks.

**Uniformitarianism.** The concept that processes now occurring in the earth also operated in the past.

**Visible light.** Electromagnetic radiation with wavelengths from $7 \times 10^{-7}$ to $4 \times 10^{-7}$ meters

**Volcanic eruptivity index (VEI).** Measure of the amount of lava and ash erupted from a volcano.

**Westerlies.** Winds that blow toward the east in temperate latitudes.

# SELECTED BIBLIOGRAPHY

## CHAPTER 1: ATMOSPHERE, OCEANS, AND RIVERS

Ambrose, S.H. (1998). Late Pleistocene human population bottlenecks, volcanic winter, and differentiation of modern humans. *Journal of Human Evolution*, 34, 623–651. Proposes that the eruption of Toba caused diversification of people to modern races.

Carson, R. (1979). *The Sea Around Us.* New York: Oxford University Press. A major book on the oceans for general readers.

Couper-Johnston, R. (2000). *El Niño: The Weather Phenomenon That Changed the World.* London: Hodder and Stoughton.

Fagan, B.M. (1987). *The Great Journey: The Peopling of Ancient America.* New York: Thames and Hudson.

Keys, D. (2000). *Catastrophe: An Investigation into the Origins of the Modern World.* New York: Ballantine Publishers. The effect of the eruption of Krakatoa in 535 AD on world history.

Kington, J. (1998). *The Weather of the 1780's over Europe.* Cambridge: Cambridge University Press.

Lamb, H.H. (1995). *Climate, History, and the Modern World.* London: Routledge.

Moorehead, A. (1962). *The Blue Nile.* New York: Harper and Row.

———. (1971). *The White Nile.* New York: Harper and Row. Two books by Alan Moorehead describe exploration and history of both branches of the Nile.

Parker, M. (2007). *Panama Fever: The Battle to Build the Canal.* London: Hutchinson.

Ryan, W., and Pitman, W. (1998). *Noah's Flood: the New Scientific Discoveries about the Event that Changed History.* New York: Simon & Schuster. Proposes that the biblical flood occurred in the Black Sea.

## CHAPTER 2: TECTONICS

Friedrich. W.L. Translated by A.R. McBirney. (2000). *Fire in the Sea: The Santorini Volcano: Natural History and the Legend of Atlantis.* Cambridge: Cambridge University Press.

Hallam, A. (1983). *Great Geological Controversies.* Oxford, New York: Oxford University Press. Covers much of the history of geology.

Morgan, W.J. (1972). Plate motions and deep mantle convection. In *Studies in Earth and Space Sciences.* R. Shagam, R.B. Hargraves, W.J. Morgan, F.B. van Houten, C.A. Burk, H.D. Holland, and L.C. Hollister, eds. *Geol. Soc. Amer. Memoir,* 132, 7–22. New York: Simon & Schuster. Outlines the concept of mantle plumes.

Nur, A., and Ron, H. (2000). Armageddon's earthquakes. In W.G. Ernst, and R.G. Coleman, eds. *Tectonic Studies of Asia and the Pacific Rim.* International Book Series, vol. 3, p. 44–53. Columbia, MD: Bellwether Publishing for the Geological Society of America.

Oreskes, N. (1999). *The Rejection of Continental Drift.* New York: Oxford University Press. Discusses objections to Wegener's proposal of continental drift.

———. *Plate Tectonics: an Insider's History of the Modern Theory of the Earth.* Boulder, CO: Westview Press.

## CHAPTER 3: EVOLUTION, CREATIONISM, AND THE LONG HISTORY OF THE EARTH

Alvarez, W. (1997). *T. Rex and the Crater of Doom.* Princeton: Princeton University Press. The discovery of the asteroid impact at the end of the Cretaceous; told by the person who made the discovery.

Darwin's complete published works and many unpublished documents are online at www.darwin-online.org.uk.

Dunsworth, H.M. (2007). *Human Origins 101.* Westport, CT: Greenwood Press.

Hallam, A. (2004). *Catastrophes and Lesser Calamities: The Causes of Mass Extinctions.* Oxford, NY: Oxford University Press.

Landing, E., Myrow, P., Benus, A., and Narbonne, G.M. (1989). The Placentian Series: Appearance of the oldest skeletized fossils in southeastern Newfoundland. *Journal of Paleontology,* v(63), 739–769. Describes a well-preserved transition from the Precambrian to the Cambrian.

Lawrence, J., and Lee, R.E. (1950). *Inherit the Wind.* A play, and later a movie, which dramatizes the Scopes trial.

Martin, P. (2005). *Twilight of the Mammoths: Ice Age Extinctions and the Rewilding of America.* Berkeley: University of California Press. About extinction of the megafauna by overhunting.

Moore, J., and Moore, R. (2006). *Evolution 101.* Westport, CT: Greenwood Press.

Repcheck, J. (2003). *The Man Who Found Time: James Hutton and the Discovery of the Earth's Antiquity.* Cambridge, MA: Perseus.

Retallack, G, Veevers, J.J., and Morante, R. (1996). Global coal gap between Permian-Triassic extinction and Middle Triassic recovery of peat-forming plants. *Geological Society of America Bulletin,* v(108), 195–207.

Wegener, A. Translated by R. von Huene. (2002). The origins of continents. *International Journal of Earth Sciences,* 91(supplement), S4–S17. This is a translation of the original paper by Alfred Wegener (1912), *Die Entstehung der Kontinente.* Geologische Rundschau, vol. 3, 276–292.

Weiner, J. (1995). *The Beak of the Finch.* New York: Random House. About Darwin's observations in the Galapagos Islands.

Winchester, S. (2001) *The Map That Changed the World: William Smith and the Birth of Modern Geology.* New York: HarperCollins.

## CHAPTER 4:  RESOURCES AND THE ENVIRONMENT

Berners-Lee, T. (2000). *Weaving the Web*. San Francisco: Harper.

Carson, R. (1962). *Silent Spring*. Boston: Houghton-Mifflin. One of the first, and most influential, environmental books.

Chrispeels, M.J., and Sadava, D.E. (1994). *Plants, Genes, and Agriculture*. Boston: Jones and Bartlett.

Crowley, D., and Heyer, P., eds. (2007). *Communication in History: Technology, Culture, Society*. Boston: Pearson Allyn & Bacon.

Deffeyes, K.S. (2005). *Beyond Oil: The View from Hubbert's Peak*. New York: Hill and Wang.

Drege, J.P., and Buhrer, E.M. (1989). *The Silk Road Saga*. New York: Facts on File.

Fagan, B.M. (2004). *The Little Ice Age: How Climate Made History, 1300–1850*. New York: Basic Books.

Holliday, J.S. (1999). *Rush for Riches: Gold Fever and the Making of California*. Berkeley: University of California Press.

Hutchinson, B. (1998). *Fool's Gold: The Making of a Global Market Fraud*. Toronto: A.A. Knopf, Canada. About the Bre-X scandal.

Kiple, K.F. (2007). *A Movable Feast*. Cambridge: Cambridge University Press. History of food from hunter-gatherers to the present.

Lappe, F.M. (1991). *Diet for a Small Planet*. New York: Ballantine Books. Discusses the problem of hunger in the world.

Rhodes, R. (1986). *The Making of the Atomic Bomb*. New York: Simon and Schuster.

Sobel, D. (1995). *Longitude: The True Story of a Lone Genius Who Solved the Greatest Scientific Problem of His Time*. New York: Walker. About the clock that made it possible to determine longitude at sea.

Yergin D. (1991). *The Prize: The Epic Quest for Oil, Money and Power*. New York: Simon and Schuster.

# Index

**About the Authors**

JOHN J.W. ROGERS is the Emeritus W.R. Kenan, Jr., Professor in Geology at the University of North Carolina. He completed he Ph.D. work at Caltech in 1954 and was hired as one of three faculty to staff the new Geology Department at Rice University. He has conducted research in the western United States, the Caribbean, North Africa, the Middle East, and India. He is a Fellow and Past President of the International Division of the Geological Society of America, a Fellow of the Geological Society of India, and an Honorary Fellow of the Geological Society of Africa. Rogers is the author, co-author, editor, or co-editor of seven previous books.

TRILEIGH TUCKER is Associate Professor and Director of the Environmental Studies Program at Seattle University. She earned her Ph.D. in Geology at the University of North Carolina at Chapel Hill, with emphasis on Precambrian geochemistry. Her research interests and publications focus on phenomenology of science, particularly in the realms of science and religion, natural history, and sustainability as they connect science and humanities.